Astronomers' Universe

For other volumes:
http://www.springer.com/series/6960

David Schultz

The Andromeda Galaxy and the Rise of Modern Astronomy

David Schultz
Hamline University
St. Paul, Minnesota, USA

ISSN 1614-659X
ISBN 978-1-4614-3048-3   ISBN 978-1-4614-3049-0 (eBook)
DOI 10.1007/978-1-4614-3049-0
Springer New York Heidelberg Dordrecht London

Library of Congress Control Number: 2012933795

© Springer Science+Business Media New York 2012
This work is subject to copyright. All rights are reserved by the Publisher, whether the whole or part of the material is concerned, specifically the rights of translation, reprinting, reuse of illustrations, recitation, broadcasting, reproduction on microfilms or in any other physical way, and transmission or information storage and retrieval, electronic adaptation, computer software, or by similar or dissimilar methodology now known or hereafter developed. Exempted from this legal reservation are brief excerpts in connection with reviews or scholarly analysis or material supplied specifically for the purpose of being entered and executed on a computer system, for exclusive use by the purchaser of the work. Duplication of this publication or parts thereof is permitted only under the provisions of the Copyright Law of the Publisher's location, in its current version, and permission for use must always be obtained from Springer. Permissions for use may be obtained through RightsLink at the Copyright Clearance Center. Violations are liable to prosecution under the respective Copyright Law.
The use of general descriptive names, registered names, trademarks, service marks, etc. in this publication does not imply, even in the absence of a specific statement, that such names are exempt from the relevant protective laws and regulations and therefore free for general use.
While the advice and information in this book are believed to be true and accurate at the date of publication, neither the authors nor the editors nor the publisher can accept any legal responsibility for any errors or omissions that may be made. The publisher makes no warranty, express or implied, with respect to the material contained herein.

Cover illustration: Andromeda Galaxy, IRAS 00400+4059, M 31, Messier 31, NGC 224. Image taken by NASA/ESA Hubble Space Telescope.

Printed on acid-free paper

ACC LIBRARY SERVICES AUSTIN, TX (www.springer.com)

*For Richard DeLuca,
my first astronomy teacher.
Thank you.*

Acknowledgments

Thanks go out to so many individuals who helped make this book possible. First thanks go to Alex Hons and Wayne Orchiston, my astronomy professors at James Cook University, as well as my classmates. All of them were patient and supportive of my research. I especially wish to acknowledge Wayne; it was under him that much of the preliminary research on the Andromeda Galaxy was done.

Before these individuals stands Richard DeLuca – my first astronomy teacher. His classes at the Roberson Planetarium in Binghamton, New York, inspired me beyond words.

But beyond my astronomy professors, I also thank many other professors across programs in philosophy and political science. As the reader shall find, this book crosses many fields of study as it examines astronomy and the Andromeda Galaxy. It is this interdisciplinary approach that in part makes this book unique and such a pleasure to have written.

The Minnesota Astronomical Society members also deserve special recognition. Their enthusiasm for astronomy is an inspiration to me.

Finally Helene my wife deserves acknowledgement for listening to me discuss the topic of the Andromeda Galaxy way more than any person should. Her patience with me is remarkable.

About the Author

David Schultz is an avid amateur astronomer with a master's degree in astronomy from James Cook University. He is a Hamline University professor in the School of Business, where he teaches classes in government ethics, public policy, and public administration. He also holds appointments at the Hamline University and Minnesota schools of law, where he teaches election law, professional responsibility, and state constitutional law. Professor Schultz is the author of more than 25 books and 80 articles. In addition to a degree in astronomy, he has a Ph.D. in political science and a law degree from the University of Minnesota, a masters of law from the University of London, a masters degree in political science from Rutgers University, a masters degree in philosophy from Binghamton University, and a bachelors degree in political science and philosophy from Binghamton University.

Contents

Acknowledgments vii
About the Author.. ix

1. The Wonder of the Andromeda Galaxy............................ 1
2. Early Depictions of Andromeda.. 19
3. A Single Closed Theory of the Universe........................... 45
4. Andromeda and the Technological
   Revolution in Astronomy 69
5. Andromeda and Astronomy at the Beginning
   of the Twentieth Century 105
6. The Andromeda Nebula and the Great
   Island-Universe Debate 135
7. Edwin Hubble, an Infinite Universe,
   and the Classification of Galaxies 157
8. Andromeda, Galactic Redshift,
   and the Big Bang Theory 181
9. Andromeda, Cosmology, and
   Post-World War II Astronomy... 207

xii   Contents

10. Astronomy and Andromeda at the
    Close of the Twentieth Century 233

11. The Andromeda Galaxy into the Twenty-First
    Century and Beyond 253

Appendix: Andromeda Facts 259

References 261

Index 267

1. The Wonder of the Andromeda Galaxy

Who we are as humans is often dependent upon how we define our position in the universe. At the center of that definition since ancient times has been the Andromeda Galaxy. The astronomical story of the Andromeda Galaxy is not simply the tale of a celestial object, a specific tool such as the telescope, or of a particular science. Instead, it is a story about humans, history, and how we view ourselves as living beings.

However, the tale of the Andromeda Galaxy is even broader; it is a story of the universe. It is the story of how humans have defined the origins and expanse of the cosmos, themselves, and their role in it. The story of the Andromeda Galaxy thus is part science, philosophy, theology, sociology, and psychology. It is the story connecting many threads of human existence, revealing the changing depictions of how we view human nature and its role in the universe. This book is an effort to tell one small part of the story of how the study and depictions of the Andromeda Galaxy have driven and been at the forefront of the history of astronomy.

Andromeda in the Sky

Think about all the objects that can be seen in the sky. With the naked eye the number is in the hundreds or thousands, while with a telescope the number is infinite. There are also many fascinating star patterns or constellations. But the constellation Pegasus is one that captures attention.

2 The Andromeda Galaxy and the Rise of Modern Astronomy

FIGURE 1.1 The constellation Pegasus.

Pegasus dominates the northeastern skies of the northern hemisphere's middle latitudes in the fall. It is near the constellations of Pisces, Aquarius, Cassiopeia, and Cygnus. It is one of the 48 constellations named by Ptolemy, an Egyptian astronomer during the Roman empire (A.D. 90–168), and it is one of the 88 modern constellations. It is a very old constellation, its name coming from the ancient Greeks who thought it resembled the fabled winged horse Pegasus (Πήγασος). It is an imposing constellation compared to many others, occupying a large part of the sky. But the most defining characteristic of Pegasus is the "Great Square," formed with its three brightest stars – Markab, Scheat, Algenib – and Alpheratz from the constellation Andromeda. The square alone is enough to attract attention, but there is something else near the constellation of interest. To the naked eye it looks like a hazy smudge located outside of the Great Square in the constellation of Andromeda.

This smudge is the Andromeda Galaxy, M31, Messier object number 31 as identified and cataloged by Charles Messier (1730–1817) in the eighteenth century. Messier was a comet hunter who created a catalog of objects he initially thought to be comets but

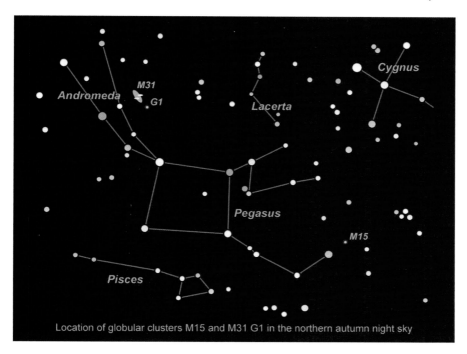

FIGURE 1.2 Pegasus constellation depicting the location of the Andromeda galaxy (M31).

which he eventually concluded were nebulae (interstellar gas) or star clusters. The Andromeda Galaxy, as it is now known, was previously and well into the early part of the twentieth century referred to as the Andromeda Nebula. Until the 1920s it was thought to be part of our galaxy the Milky Way. In fact, everything in the universe was thought to be part of the Milky Way. It was assumed by many modern astronomers that the universe consisted of one galaxy, which was ours. The Andromeda Nebula, the smudge in the sky, was simply a part of our own galaxy.

Edwin Hubble in the 1920s demolished that theory, establishing it as a distinct island in the broader universe. In so doing, he demonstrated a universe much larger than previously thought, with Andromeda even further away than envisioned. Today astronomers calculate the distance from Earth and the Milky Way to Andromeda to be about 2.5 million light-years. Andromeda is the most distant object the naked eye can see. What we see tonight is light that began traveling 2.5 million years ago. Put into perspective, the light one sees tonight from Andromeda is more than one million years older than the human race (*Homo sapiens*). Light leaving the galaxy today may reach Earth long after humans are no longer on this planet.

Prior to Messier and modern astronomy, there were many myths and beliefs regarding what this smudge was thought to be. The different accounts of what this smudge is, how studying it changed astronomy, and how it affected how humans think about the universe is the central story of this book. Looking out at the sky to see Andromeda, be it with the naked eye, a telescope, radio dish, or the Hubble Telescope, has repeatedly been a practice of astronomers from the earliest days of science. Andromeda inspires wonder.

Astronomy and Wonder

Written on the tombstone of Immanuel Kant (1724–1804), a famous philosopher and cosmologist, is the statement: "Two things fill the mind with ever new and increasing wonder and awe: the starry heavens above me and the moral law within me." According to Kant, there is a connection between the heavens and the soul. Most people never think of the two as related, but they are. Go out some clear night and look at the sky at night. Perhaps the Moon is out, or maybe it is a dark moonless night. Stare at the stars. We see star patterns, items that look like bears (Ursa Major), warriors (Orion), or other objects such as the Big and Little Dippers, the letter "W" Cassiopeia, scorpions (Scorpio), and perhaps even imaginary objects, such as winged horses, the constellation Pegasus, near where the Andromeda Galaxy is visually located. We also see objects of varying color and brightness; some seem to move in the sky over a period of nights, such as the planets. Some such as meteors appear to flicker, brighten, and then disappear in a matter of seconds; others last longer, such as comets, and other phenomena, while other residents, such as the Milky Way, serve as a mysterious permanent haze in the sky.

It does not really matter what we stare at. The reaction for many is the same – wonder. We look at the sky and wonder what

FIGURE 1.3 Meteor streaking across the sky.

it all means. Why do the stars seem to be arrayed in patterns, why are there different colored stars, how far away are these objects, and where did they come from? Similar questions may be asked of the Moon and perhaps even of the Sun. Moreover, sitting long enough and thinking about them, one might begin to ask even deeper questions. These may include whether others are also looking at these objects, are there other planets like Earth around these other stars, and if so, is there life elsewhere in the universe and perhaps could there not be another being on another planet somewhere else looking up at the sky wondering and asking similar questions?

Staring up at the sky also prompts ethical and theological questions. We begin to think about the vast sky and how small we are, perhaps concluding that we are merely a small speck in the cosmos. We wonder who created the universe, when, why, and for what purpose. Or perhaps we think, as was the case for eons, that humans occupy a central role in the universe. We are the master species on Earth, located in the center of a cosmos, and everything revolves around us. Thus, we look up to the sky, see the cosmos, and then eventually turn back to our souls and ourselves and ask about human existence. Looking outward forces us to look inward. This is the significance of Kant's tombstone quote – he saw the relationship between the cosmos and who we are. This is what many of us feel when we stare at the sky, and it is a common experience that goes back thousands of years.

Aristotle, an ancient Greek philosopher, once said that all of philosophy starts in wonder. Ancient humans, too, wondered and marveled when they looked at the sky – one even brighter and more imposing than the one we now see under the glare of city lights and pollution.

Thales, another ancient Greek philosopher, was reputed to have fallen into a well while walking and gazing at the sky. He was deep in thought and wonder. Socrates (469–399 BCE), perhaps one of the first great ancient Greek philosophers, is credited with

FIGURE 1.4 Aristotle (384–322 BCE).

The Wonder of the Andromeda Galaxy 7

FIGURE 1.5 Thales (624–546 BCE).

FIGURE 1.6 Socrates.

turning his observations from the sky to ethics. Astronomy gave birth to ethics and politics. Socrates took the concept of wonder that was applied externally to the sky and turned it inward, seeking to ask questions about human nature. The vast macrocosm of the universe provides a key to understanding the mysterious microcosm of the human psyche. Socrates was part of the original group of thinkers, both in the European west and elsewhere in the world, who tried to understand and explain the origins of the universe and how humans fit within it.

Construction of human and individual identity is relational, determined by our connection to many other factors. We define our identity socially compared to other individuals, cultures, and nations. We also define ourselves in contrast to other species, seeing *Homo sapiens* at the apex of an animal kingdom, both connected to and separate from other animals. But humans also define themselves cosmologically, constructing an identity in relationship to our role and location within the universe. Perhaps one constant of human nature from ancient times, then, has been the staring up at the sky and wondering.

Explaining the Origins of the Universe

This wondering has driven explanations of the universe. Today the prevailing theory is that the universe was created out of a Big Bang approximately 13.7 billion years ago. But for the ancient Greeks, several theories were developed to explain the cosmos.

Early Greek concepts of the world and the cosmos approached the world in what Henri Frankfort described as an "it-thou" relationship. The universe and the world were seen as living entities with animal and perhaps human tendencies or characteristics. Explanations of the rising and the setting of the Sun, changes in the sky, and other natural phenomena were seen as caused by forces by the gods or other animate forces. This stands in contrast to images of modern science that depict the universe in mechanistic ways. Thus, myths were critical to explanations. But the pre-Socratic philosophers, those writing and wondering before Socrates, began moving away from myth and toward alternative explanations. It was the beginning of efforts to engage in scientific explanations.

Hesiod's *Theogony* describes an era when there was no differentiation between heaven and Earth. All was one. But heaven and Earth were male and female, and sexual imagery was used to detail the breakup of a primordial unity and creation. A world order emerged where there were heavens above, Earth situated within an ocean, and then a Tartarus or underworld below. From this world order the struggles among Zeus and other Olympian gods was employed to explain the rise of humans, the division in cultures, and class and other social structures that Hesiod and other Greeks saw around them.

Anaximander, too, assumed a time when an infinite existed. He agreed with Hesiod that this infinite was an undifferentiated unity of the heavens and Earth. But unlike Hesiod, who used sexual imagery to explain the separation, Anaximander assumed the existence of opposites as driving creation. These opposites, hot cold, dry, and moist, are driven to separate by some motions or movements that lead to their distinction. Eventually these four opposites would later become described as four basic elements – Earth, air, fire, and water – perhaps the first effort at constructing or defining the building blocks or elements of the universe. Their separation is not explained in anthropomorphic ways but instead mechanistic, and the product of some primordial forces or vortices

FIGURE 1.7 Anaximander (610–546 BCE).

of change. Once separated, fire traveled to the edges of the cosmos to form the sky, Sun, and heavens. Air stands between the fire and Earth, with the latter sitting in and surrounded by water. Out of this vortex eventually animals and creatures emerge from the moist or the water, eventually producing other animals and then humans.

Anaximander thus uses opposites, or a clash among them, to explain change, the origins of the universe, and the rise of humans as a species and along with their culture. But his theory or origins also had a theory of finality; the vortex would someday end, the opposites would collapse, and all would revert to an infinite once again. Within his theory then are elements of astronomy, cosmology, and a little bit of evolutionary theory sounding similar to accounts later offered by Charles Darwin in the nineteenth century. His theory of an origin sounds like the Big Bang, and his end like the Big Crunch, a term astronomers employ to describe the end of the universe when it begins to contract after it has stretched out as far as possible from the original explosion. But Anaximander's reliance on oppositions and their clash to explain change was a common feature of the ancient Greeks, as well as other cultures. The Chinese, for example, relied on Ying and Yang to describe cosmic change and the human struggle.

Other ancient Greeks offered other explanations of the universe and change. Anaximenes saw air as the primordial element of the universe – its basic building block. Compression and dilation of air explains the origin of the universe out of a great infinite. Xenophanes saw water as the core element, warmed or cooled to form the universe.

Democritus described the basic elements of the universe as composed of invisible small particles – atoms, or "atomos" in ancient Greek, which meant indivisible. Their motion, interaction, and combination form the core of the world. He proposed that were one to cut up something infinitely small, such as a loaf of bread, there would eventually reach a point beyond which cutting could not occur. A basic particle existed out of which all other matter or compound entities formed. From these elementary particles we get a universe, humans, and all that we see around us.

Other ancient Greek philosophers had their theories about the universe. Pythagoras thought numbers to be real entities and

FIGURE 1.8 Democritus (460–371 BCE).

FIGURE 1.9 Pythagoras (circa 570–490 BCE).

saw them as the basis of the world (a concept not too distant in many ways from theoretical physics based on math).

Pythagoras's views influenced Plato, who eventually described a theory of the forms as reality. What we saw around us was temporal and transitory and of lesser reality than another realm of truth

FIGURE 1.10 Plato (429–347 BCE).

and permanence. Behind all of what we see in the world, from the cave and shadows as he would describe in *The Republic*, resided a permanent world of forms and true knowledge. The world we inhabited was a faint resemblance of this world, and the task of the philosopher was to grasp this world. Our senses tricked us, and we needed science or a method of inquiry to find true reality and knowledge.

Plato's quest for true knowledge built on the pre-Socratics. All of them shared some common characteristics. First, they all sought to explain the origins of the universe, Earth, and humans. All of them were gripped with answering the most basic questions of science, philosophy, and theology. Why is there something rather than nothing? Why does the universe exist? Who created it, and for what purpose? Where did Earth come from? How did humans come about, are we alone, and what role is there in the universe for us? The basic question of science, explaining why something as opposed to nothing, led to important questions about ethics, politics, and religion. Describing a purposive universe created by a God or gods, for example, suggested something about who we were as persons. Moreover, if humans were crafted in the image of a deity, as the Greeks, Romans, Christians, and others thought, that also suggested something about the relationship of our species to other animals, and of individual humans to others. If we were created in God's image on Earth, should we not have dominion over them as suggested in the biblical book Genesis?

Another common trait or feature of these early writers was the effort to explain change. Stare at the sky and there are many common and constant features. Stars continue to twinkle, constellations remain fixed (at least over periods of time longer than a human life), and the haze of the Milky Way does not change. But some stars (actually, planets) move across the sky. Comets appear; some stars suddenly brighten or fade. The Sun rises and sets, and the Moon moves across the sky and changes its relative position on a monthly basis. How do we explain this change? Do these changes portend anything?

For the ancients, the study of the sky and these changes did prophesize events, thus the field and rise of astrology. Human lives and fates were connected to the stars. Change versus constancy was a dominate theme of the ancient Greeks.

FIGURE 1.11 Heraclitus (circa 500 BCE).

Heraclitus, a Greek philosopher, once stated one cannot step in the same stream twice. Perhaps more correctly he should have said one cannot step in the same stream once. There is no permanence, only flux. Parmenides and Zeno, two other ancient Greek philosophers from the same era, made similar arguments. The world was one of change, but why it changed was a matter of speculation. They built their speculation on the works of Anaximander and Anaximenes, for example, asking how from some infinite or primordial unity or beginning something changed to produce the world around us. The questions they asked eventually forced Plato to propose a dualism – a world of change we see and a world of permanence that resides in ideas and the intellect. Later, Aristotle, a student of Plato, argued for some first or final cause of the universe, perhaps a god.

A third common theme of the ancient Greeks was connecting the microcosm to the macrocosm. By that, all sought to explain the universe by appeal to smaller units. The universe is made up of basic elements, or Earth, air, fire, and water. The universe is composed of atoms, or of numbers, or of forms. There is a connection between the vast cosmos and small particles.

Identifying the basic particle, unit, or building block of the universe remains a common concern of physics and astronomy today, with scientists now in search of the Higgs boson particle. According to Nobel Prize-winning physicist Leon Lederman, the Higgs boson is the über-particle, or the God particle in pop culture; it is the primordial unit that is the building unit for atoms, protons, neutrons, electrons, and all the other subatomic particles discovered in the last 50 years.

Linking the macrocosm to the microcosm or vice versa was used not only to explain the construction of the universe but the relationship of human nature to politics. In seeking to explain what justice is, Plato in his *Republic* analogized that we learn about justice in the soul by first looking at what is justice in the state. There is an intimate linkage between the two. In fact, Plato would describe in the *Republic* how there are different types of humans – men of gold, silver, and brass – and how their characters lead to different capacities and talents, such as who are more suited to rule or be ruled. Finally, both in the *Republic* and then in some of his other writings such as the *Timaeus*, Plato would connect the state to a broader theory of reality, to a conception of the universe, the gods, and a creator, and seem to argue that the order of things on Earth, in governments, and perhaps in human nature, mirror or align with one another. There is a connection between the cosmos and who we are and the lives we live. Humans have a place in the cosmos.

The Great Chain of Being

This cosmology did not end with the Greeks but has persisted throughout Western history at the least. Christians, especially in Genesis, described a world or universe created by a God in 7 days. Humans were forged out of the Earth and created in the deity's image, and a duality of heaven and Earth was used to account for the temporal, permanence, and change. Arthur Lovejoy, a famous twentieth-century historian of ideas, once argued that Western science and theology was premised upon a "great chain of being." It was a chain that stretched from God to the angels and the heavens, to the stars, Earth, humans, animals, and then to hell and Satan.

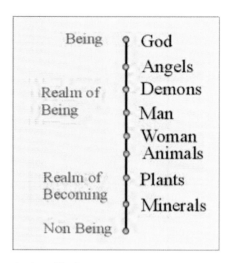

FIGURE 1.12 Great chain of being.

The Great Chain of Being was a descending hierarchy of reality, of goodness, and authority. It was meant to describe the order of the theological universe, define political authority and who was fit to rule, and also to explain good, evil, freedom, and immorality. It was an all encompassing theory. It was a theory that justified papal authority as coming from God and therefore as lording over or ruling over secular governments in medieval Europe.

However, it was a theory premised upon a cosmology. The Church assumed Earth to be at the center of a universe with the pope at the center of Earth. The entire universe revolved around a geocentric Earth that was permanent and fixed. The authority of the Church and the pope was directly and vertically descended from a god. Challenge this cosmology and one defies the Church. This was what Copernicus did in 1543, when he contended that the Sun and not Earth was the center of the Solar System. The same was true for Galileo.

Ptolemy, in ancient Egypt, as well as the Greeks, Romans, and Christian Europeans, all had assumed that Earth was at the center of the universe. Yet to hold this belief, astronomers had to make a series of assumptions about planetary movements, including that the planets seem to occasionally move backwards or in small circles as they revolve around Earth. Increasingly, these models became more complex and unable to predict accurately the movement of objects in the sky. Copernicus sought to develop a model to overcome these errors. He thus speculated in his book: "Although there are so many authorities for saying that the Earth rests in the center of the world that people think the contrary supposition is ridiculous and inopinable; if, however, we consider the thing attentively, we will see that the question has not yet been decided and is by no means to be scorned."

Copernicus thus speculates on the implications of a heliocentric or Sun-centered conception of the universe. His theory, as proposed in his posthumous 1543 *On the Revolution of the Heavenly Spheres*, proved superior to the geocentric models. In 1616, the pope declared the book heresy and forbade Catholics to read it. Why was the book denounced? A Sun-centered model of the universe threatened the foundations of the Christian order. Medieval Christian theology and the Church had built its entire edifice upon the assumption that Earth was the center of the universe and not merely one of several equal planets. The geocentric model defined man's and woman's relationship to God and one another, and, in the process, defined who humans were. What the Church had done was to create a great theological chain of being, with Earth lying in the central position and the Catholic Church at the center of the chain; and the pope, deriving his authority from God, was thus able to claim he had a privileged position to lord over the Earthly empire.

In decentering Earth from the apex of the universe, Copernicus did not just challenge scientific knowledge; he changed everything and threatened the order of the world. Galileo did that too in 1610 when he observed Jupiter's moons revolve around it and suggested Earth did the same with the Sun. Johannes Kepler's laws of planetary motion devised in the early seventeenth century showed planetary motion not to be perfect circles but ellipses, again casting a blemish upon the supposed perfection of a God-directed universe. Similarly, Isaac Newton's 1687 *Principia Mathematica* proposed laws of motion and gravity that replaced God with mechanistic forces of attraction. Then, in the twentieth century, Albert Einstein's special theory of relativity in 1905 would question whether in fact there is a center of the universe, and theories of the Big Bang poked holes in whether God created heaven and Earth in 7 days. Science challenged cosmology but it also challenged religion, politics, and how we think about human nature. As Friedrich Nietzsche, a nineteenth-century German philosopher once exclaimed, since Copernicus humans have been slipping further and further away from the center and perhaps into nothingness.

Each step in the process of adding to scientific knowledge forces changes in knowledge about the universe and with that, a corresponding change in human affairs. New scientific knowledge challenged old truths, revealing them to be nothing more than dogma.

What Does All This Have to Do with the Andromeda Galaxy?

This brief story of science, theology, and politics brings us back to Kant and then the Andromeda Galaxy again. It seeks to establish the linkage between the heavens or the sky and what we as individuals believe about ourselves and our lives. Scientific reason can construct models of the universe, but it can also destroy old beliefs and the moral laws we hold. Thus, how we view the universe is important to how we view ourselves.

The Andromeda Galaxy is part of the story of the cosmos. As an object of curiosity in the sky it has attracted attention since the first time humans gazed upwards. Humans have continuously wondered what it is. Whenever some new technology emerged, humans turned to look at it, whether it is with a telescope, a spectroscope, a radio dish, or with the Hubble telescope. Astronomers have also turned to M31 and observed it in ultraviolet and infrared light as the technologies for observing in these frequencies have evolved. Moreover, as we have come to learn, and as this book will tell, since the Andromeda and Milky Way galaxies are neighbors and resemble one another in many ways, scientists have turned to the former to learn more about the latter. The Andromeda Galaxy has provided rich data and observations that have been important to the advancement of astronomy and our knowledge of the universe. Astronomer Paul Hodge states it well: "M31 forms an important testing ground for ideas about massive galaxies and about galaxy evolution, and is ripe for detailed astrophysical exploration."[1] As Andromeda has been observed, its study has been the story of the rise of modern astronomy. What we have learned about it has not only driven the field of astronomy and the study of the cosmos, but also of our view of the role of humans in it.

---

[1] Paul W. Hodge, *Atlas of the Andromeda Galaxy*, Seattle: University of Washington Press, 1981, p. 11.

2. Early Depictions of Andromeda

Introduction

In ancient Greek mythology Andromeda is the daughter of Cepheus, the Ethiopian king of Joppa, and Cassiopeia. Cassiopeia angers Poseidon by claiming Andromeda is more beautiful than the Nereids (daughters of Poseidon). In retaliation, Poseidon sends a monster to prey upon Ethiopia, with the sacrifice of Andromeda required to halt this action. Andromeda is chained to a rock by the sea and is eventually rescued by Perseus, who kills the monster and marries Andromeda. Her final fate, along with Cepheus and Cassiopeia, is to be carried up into the sky as constellations; that is why all of them hover over our heads for eternity.

Although Andromeda's placement in the sky may be explained by her defiance of the gods in Greek mythology, Messier 31 has occupied a central but often overlooked and underappreciated place in astronomy. In modern astrophysics M31 is recognized as the Milky Way's closest spiral cousin, though this has not always been true. In fact, until the 1920s and Edwin Hubble's landmark research on stellar distances, the Andromeda Galaxy was seen as simply another nebula located well within the confines of our galaxy. Efforts to define what M31 is, to measure its distance, and even now, the study of it, have provided powerful clues to the nature of the Milky Way as well as to a cosmological understanding of the origins of the universe. The changing descriptions of Andromeda thus parallel and provide insights into the history of astronomy.

However, given Andromeda's central role in astronomy – and the lure it has had upon astronomers – it is surprising how little has been written that describes its unique role in history. Many scientific papers have discussed various aspects of the Andromeda Galaxy, but few works, with the notable exception perhaps

20 The Andromeda Galaxy and the Rise of Modern Astronomy

FIGURE 2.1 Andromeda chained, as depicted on an ancient Greek vase.

of those by Paul Hodge (1992), have sought to fill this void. This chapter describes the ancient and early modern (up until the nineteenth century) history of the Andromeda Galaxy in the field of astronomical research.

Ancient Astronomy

Ancient depictions and explanations of astronomical phenomena dramatically contrast with those offered by modern astronomy. These ancient explanations were part myth and part theology. In those days, the sky was often described as the playground of the gods. Many of the star patterns across cultures referred to the constellations as gods. Or, more simply, the brightest stars or planets were often assumed to be gods – Mercury, Venus, Mars, Jupiter, and Saturn were all gods for the Romans. Even to this day, astronomical convention is to use the name of mythic gods and goddesses from various cultures to name moons and many other objects in the sky. Additionally, the causes of everyday astronomical events, such as the rising and setting of the Sun, were seen as caused by a god. So were special events, such as the appearance of new stars, comets, perhaps the conjunction of planets.

However, natural phenomena were also depicted in anthropomorphic terms. Human attributes were ascribed to natural phenomena to explain the rise and origin of the universe. Henri Frankfort described this contrasting relationship of the ancient Greeks and others to their universe as one pitting myth against science.

Ancient cultures described the world as a "thou," seeing it as a living person-like entity. Change was the product oftentimes of male-female interactions, rendering explanations of change or the origins of the universe in some type of sexual imagery or in some way directed by human-like properties. Pre-scientific societies and cultures saw their relationship to the world in mythic and personal ways. While causal (cause and effect) scientific explanation today is premised upon theories of forces and mechanics, thanks to Newton, Einstein, and the laws of particle physics, for example, pre-scientific societies often used personal explanations to account for phenomena such as the rising and setting of the Sun, or the appearance of constellations. Attributing change in the universe or the world to human-like or animalistic type activities made it easy to explain natural phenomena in that one could say that the pattern of the stars in the sky was intentional, or that the appearance of comets, meteors, or the movement of the planets were part of some purposeful design. In fact, one could go so far as to argue that the universe itself had a purpose, that it was designed by someone for some goal. If everything else we observed had a maker, so must the universe, too. In contemporary pop culture, such as Douglas Adam's *Hitchhiker's Guide to the Galaxy*, the number "42" reveals the meaning of a universe that operates like a big organism. In contemporary science, the Gaia thesis depicts the universe as a living organism, although not in anthropomorphic terms.

Nature and the universe did not just exist, it had a purpose. Aristotle, in discussing the various ways to describe phenomena, stated that there were four "causes" (αἴτιον) – material, formal, efficient, and final. The material cause was the matter out of which something was made. The material cause of a chair might be wood or stone; for the universe, it could be something like Earth, air, wind, or fire, as many of the ancient Greeks thought. The formal cause referred to the shape of any object. The efficient cause referred to the force or reason why something is in motion or at rest. If some object, such as a planet, is moving across the sky, there must be some efficient cause explaining this movement. The same is true for the movement of the Moon or the Sun across the sky, the appearance and disappearance of comets and meteors, and everything else. Conversely, there must also be an efficient cause for why some objects do not move or change. Why do the sky and star patterns remain fixed? Efficient causes explain how things came about, change, or perhaps remain the way they are.

Lastly, there is the concept of a final cause. This is the purpose for something. A final cause for an egg might be a chicken, for a seed, a tree. Everything for Aristotle had a final cause. Everything was either in a state of being – it had reached it final cause – or it was in a state of becoming – moving from where it was to securing its purpose. Change was the product or process of entities moving from becoming to being.

For Aristotle, understanding the four causes of entities meant one would have knowledge about it. Applying these four causes to the universe, Aristotle followed in the line of other ancient Greek thinkers in an effort to explain it. However, of the four causes, the final and efficient causes are the most relevant to Aristotle's cosmology. If the efficient cause referred to something that brought an entity into being or existence, one had to ask what it was that accounted for the origin of the universe. Here Aristotle states that everything has a cause, with each cause preceded by another one. A chair has its cause in a carpenter who secured the wood from a tree that found its origin in a seed from another tree, which in turn found its origin in a previous tree, ad infinitum.

At some point there is a regression back so far that there is a first cause, an unmoved mover, that is the basis of all the matter and motion in the universe. This unmoved mover, at least as adopted by Christian interpretations, is God. Aristotle describes the universe as one created by the gods, with the patterns in the sky, the movements of plants, and all other phenomena as set by the gods. Aristotle describes Earth as a sphere located at the center of an eternal universe. Earth is one of the four basic elements, with the water, air, and fire above Earth's surface, located in their assigned or appropriate locations.

But the concept of the unmoved mover as the first efficient cause also connects to the idea of a final cause or purpose. If there is in fact a creator for the universe and nature then there is also purpose for the creation. The creator has a purpose for the design. To understand nature and the universe then meant that one also had to comprehend its purpose. The universe was not simply some inanimate blob of material operating according to impersonal mechanistic forces. It had a real purpose, and to understand it meant grasping this final purpose. Thus, the universe had some inner values or goals to it. For Christians, this purpose was entwined with theology. God was at the center of a universe, the great chain of being described in Chap. 1, and history, science, and all of the universe reflected a divine plan of its creator. Science was not divorced from theology but imbued with it.

The significance of this discussion is that the ancient Greeks, along with many other early cultures, all struggled with the mythic versus the scientific in seeking to describe the world. For the Ancients, the Nile circumscribed their world and connected to the heavens and the Milky Way. The Native Americans of North America had their myths about the constellations and what star shapes signified, depicting deer, wolves, bears, and other creatures of nature they confronted. The point here is that the study of the sky was done often in a pre-scientific fashion, invoking myths, anthropomorphic creatures, or deities to explain the change of seasons or other events.

Much of early astronomy was practical-seeking to explain the sky so as to determine the seasons of the year and when rains would come, and therefore when to plant or harvest. Beyond wonder, as discussed in Chap. 1, astronomy was born out of a practical desire to predict the future, for meteorological or astrological purposes. Cultures across the world built structures, including the Mayan temples of Mexico and perhaps even Stonehenge built by the

FIGURE 2.2 Mayan temple in Mexico.

Druids in England that had astronomical purposes in addition to serving other functions. The Three Wise Men of the biblical story of Jesus, too, looked to the sky and found a star that foretold of an important future event; this also demonstrated the importance the stars had in ancient cultures.

The Greeks, as noted, struggled between a pre-scientific and a slowly emerging scientific viewpoint on the world. What is meant by a scientific point of view? It involves a depiction of nature not in anthropomorphic terms but more in terms of entities governed by the rules of physics. A scientific point of view makes empirical claims that can be tested by the gathering of evidence instead of resting upon analogies, myths, or faith. It also includes claims subject to strict methods of inquiry that can be replicated. Finally, a scientific point of view generally is separated from theological or other ethical claims about purpose. A scientific explanation regarding the movement of planets in the sky would not say that they do move that way in order to fulfill some goal, such as to reveal God's plan for the eternity. Instead, the movement of planets from a scientific point of view may be described according to the laws of motion as articulated by Kepler or Newton.

The significance of this difference between a pre- or non-scientific and a scientific point of view is critical to astronomy and discussions about the universe. Although in the ancient

Early Depictions of Andromeda 25

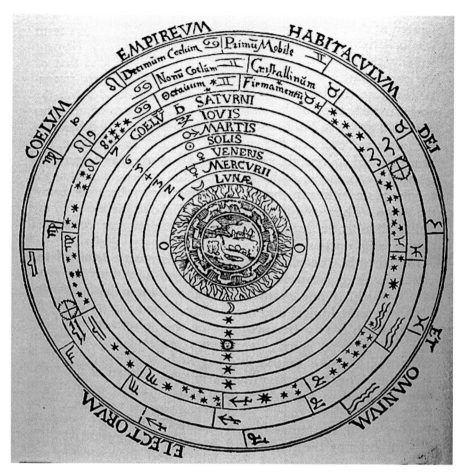

FIGURE 2.3 Medieval depiction of Ptolemy's universe.

Greek world there was movement toward a scientific worldview, that movement disappeared with the rise of Christianity and the disappearance of many of the classical texts of the Greeks. During the early Middle Ages until approximately A.D. 1200 the Christian world view was one that submerged science beneath religious faith. It synthesized some classical Greek concepts or writings but subordinated them to a view of the universe that assumed God had created and directed it according to the rules of the Scriptures. It was a world, as noted earlier, that placed God at the head of the universe as the unmoved mover or first efficient cause. It made the Earth the center of a static, finite cosmos, with the stars, planets, and all other celestial objects fixed in their place according to a plan and design by God as revealed in Genesis and in the Bible. God also served as the final cause or purpose of the universe, directing all events to reveal ultimately some purpose of His. The cosmos was a closed, finite world.

Scientific explanations of the world took a back seat to theology. Instead of reason or empirical facts determining truth, empirical observations took a back seat to faith. For example, efforts to explain the motions of the planets in the sky in a geocentric universe were complex. The basic model of the universe was premised upon a model constructed by Ptolemy, a second century astronomer, who wrote the *Almagest*. His model of the Solar System presupposed that Earth was at the center of a universe where all other objects moved in perfect circles around it.

Yet such a model could not account for certain phenomena. Specifically, how does one account for the planets moving in the sky compared to the other stars? More importantly, how can one account for the fact that planets, such as Mars, appear to stop and then appear to move backwards? Today this is accounted for by the fact that Earth and Mars revolve around the Dun in different orbits with different distances from the Sun and different speeds. At some point, as both planets revolve around the Sun, Earth passes Mars and this results in the appearance of the latter changing direction and moving in a retrograde motion.

Yet Ptolemy could not invoke this explanation since he assumed Earth was the center of the Solar System and the universe. In fact, except most notably for the ancient Greek philosopher Aristarchus (310–230 B.C.), the dominant assumption was for a geocentric cosmos. Aristarchus had proposed a heliocentric model. So for Ptolemy to account for the retrograde motion he also proposed that along with the circular movement of the heavens around Earth there were also epicycles, or small circles that the planets moved along. Thus, at certain points in their orbits the planets would literally shift direction and then move in a small circle that took them backwards. Such a movement seems inconceivable now, but given the premise of a geocentric cosmos reinforced by a Christian theology, such movement was taken as a fact.

Reality was forced to conform to theory or belief. Moreover, the movement along circles and epicycles seemed to conform to some empirical observations in that they adequately described what was being visually observed. This view of the universe persisted until Copernicus, with astronomers prior to him employing more and more complex models laden with epicycles to account for motion and movement in the sky.

This at least was what occurred in the Christian west. Knowledge and learning was different in the Islamic world. The Medieval Christian world was not the only place to face the challenge of reconciling science, religion, and philosophy. The Arabic Islamic world of the twelfth and thirteenth centuries faced similar problems. One school of Islam, located in Alexandria, Egypt, was particularly interested in these issues, and its scholars increasing turned to classical texts besides the *Koran* for inspiration and answers, and they also developed extensive commentaries on these books. Among the works that they commented on included those by Plato, Aristotle, and Ptolemy.

In effect, the writings of the Greeks and Romans, including Plato and especially Aristotle, had disappeared from the Christian Medieval world from the fifth century until the thirteenth century. The loss of these writings meant a loss of the secular scientific work of the ancients, resulting in scripture and faith dominating and directing scientific inquiry.

During the time when Europe had no access to the classical writers, however, the Arab world had an important and lively intellectual culture, of which Aristotle and Plato were a part. In fact, Arab thinkers such as Alfarabi (870–950), Avicenna (980–1037), and Averroës (1126–1198) studied the classics in detail, and it was from the Arab world, through Moorish Spain, in the thirteenth century, that Aristotle was reintroduced to Christian thinkers. Moreover, many of the changes that the Christian west would undergo as a result of the rediscovery of the classical writers was both anticipated and made possible by the Islamic world's importation of Aristotle and Plato into the latter's theological tradition.

Aristotle's works, long preserved in the Arab world, began its appearance in the twelfth century through translations by Bishop

Raymond of Toledo and William of Moerbeke. Aristotle's vision of human nature, views on government, and most importantly, his claims about nature, science, and the use of reason contrasted sharply with the Christian world based on faith. The significance of the reintroduction of Aristotle and efforts to reconcile Christianity with it was dramatic. Classical rationalism provided an alternative view of the world; it led to a questioning of many of the assumptions Christians held for centuries.

One way this rediscovery of Aristotle was important can be seen in writings of the English friar William of Ockham (1288–1348). William of Ockham (Occam) wrote and commented extensively on Aristotle, including the latter's book *Physics*. He argued for parsimony or simplicity in explanations, preferring the less to the more complex theory or arguments to explain phenomena. This appeal to simplicity came to be known as Occam's Razor. There are various ways to assert or describe Occam's Razor, but one formulation simply states: "Explanations should minimize assumptions and one should prefer the less to the more complex arguments to describe and explain things." Or perhaps, in the spirit of Occam, one could state: "Prefer the simple to the complex" or "less is more."

Occam's Razor is a major operative feature of modern science. It discourages the use of complex explanations in preference for the more simple ones. If something can be explained with fewer assumptions and assertions, we should prefer it to those that invoke the more complex. Occam's Razor is not just a contemporary tool or assumption of modern science; its import was felt already in William of Occam's time. Specifically, think of the Ptolemaic geocentric world of celestial orbits and, more importantly, epicycles. To explain heavenly movement of the planets more and more complex models and use of epicycles were invoked. The need to do this was a consequence both of increased precision of measurements (as a result of new astronomical devices) and the recurrent errors in prediction of astronomical phenomena. Thus, the movement to provide more accurate predictions led astronomers including Johannes, Kepler, and Nicholas Copernicus to reformulate and rethink many scientific assumptions held by Christian Europe.

The Christian Universe

The universe of the Christians was one created and ruled by God, according to the features and commands described in Genesis, with Earth at the center and humans (especially men) occupying the role as the master species and gender of the planet. But the development of modern science in the later Middle Ages and the Enlightenment challenged the sovereignty of faith as the sole or primary way to understand and order the world. In its place, thinkers such as Francis Bacon and Rene Descartes argued that reason, especially scientific reason, guided by certain rules or methods, could check human errors and dogmas and result in greater understanding.

Moreover, the new science challenged how the universe came to be redefined. Christians described the universe as one created and actively governed by God, with all motion and activity within it striving toward the deity. Philosophers such as Nicholas Malebranche (1638–1715) contended that the active "Vision of God" explained motion in the universe, including how the mind and body interact. This Christian universe was not simply composed of things or matter but was intertwined with moral properties of goodness because it was a creation of God. This meant that to describe the cosmos was also to ascribe moral goodness to it. These moral properties defined standards of conduct for man, through God's and natural law, and determined the best type of human laws and governments.

Scientific thinking, as it had developed to the sixteenth century, thus bore the characteristics of Christianity, Plato, and Aristotle. From Christianity, God was at the center of the explanation of why the planets, the Sun, Moon, and stars rotated around Earth, and God provided the explanation of why there was something rather than nothing. From Plato came the idea of dual reality and the concept of a demiurge. From Aristotle the universe was described in terms of gradations of reality and the four causes. If the universe could be understood as movement from becoming to being, matter could be described both as being in motion from a lesser to a greater state of reality. The universe was also purposive, not mechanical, with all entities seeking their final purposes.

What if God did not exist or what if the universe was not purposive but instead operated like a machine? Could one envision a type of science that stripped God and final causes out of it?

Scientists in the sixteenth through the eighteenth centuries challenged this vision of the universe, self-consciously seeking a break from the ways of the ancients and Christians. As early as the fifteenth century, Nicholas of Cusa (1401–1464) had contended that the universe was different from what Christians thought. He had questioned whether Earth was at the midpoint, and whether the universe had a center or finite boundaries at all. But it was in the sixteenth century that Nicholas Copernicus first questioned the belief that Earth was the center of the universe. Francis Bacon labeled religion as a superstition of the mind, challenged the relationship between faith and reason, and questioned the idea that the universe was governed by final causes. Descartes continued this line of attack by displacing God from the center of knowledge and turning toward man. And Newton completed the redefinition by recasting the universe in terms of mechanistic forces and laws of motion.

Nicholas Copernicus

Nicholas Copernicus (1473–1543) was a Polish astronomer who rebelled against the values of the ancient and Christian world and led a revolution that would help create the modern world. Copernicus tried to perfect mathematically the Ptolemaic geocentric model but found it increasingly difficult to do. Thus, initially as a speculative device, he assumed the universe to be heliocentric and not geocentric. Copernicus then explored the implications of assuming that the Sun is the center of the universe and that Earth is the third planet from the Sun, orbiting around it along with other planets. His final product, *The Revolution of Heavenly Spheres*, published at his death in 1543, assumed a different world from the Christians. His heliocentric model dispensed with epicycles, and it seemed to offer more predictive accuracy that the current one. Occam's Razor applied to astronomy suggested a contrary view of the universe than had been in place for 1,500 years.

How revolutionary was his assumption or model? In 1616, the Pope declared the book heresy and forbade Catholics to read it.

Why was it condemned? It raised many difficult questions for Christians. If Earth were not at the center of the universe this planet was merely one of several others. Could there then not be other planets with other individuals on it? If men were not the sole master race but others existed elsewhere, how might that question theological stories about Adam and Eve, for example? How might it also question the Great Chain of Being that held the Christian world together? How might it question closely held premises, such as theories of gravity that asserted that objects fall toward Earth because it was at the center of the universe?

Medieval Christian theology had built its entire edifice upon the assumption that Earth was the center of the universe and not merely one of several planets. The geocentric model defined humanity's relationship to God and, in the process, defined who we were. What the Church had done was to create a great theological chain of being, with Earth lying in the central position. The Catholic Church was at the center of the chain, and the pope, deriving his authority from God, was thus able to claim he had a privileged position to lord over the Earthly empire.

To challenge the geocentric model was to displace the pope's and the Church's central position, and to question Christian hierarchy, order, and status as dramatically as Luther and Calvin had done in regards to doctrine. Copernicus creates, in effect, a new center (of power), and the pope is not in it.

Copernicus ultimately forced a rethinking of the universe, away from one that was God-directed to one run differently somehow. Perhaps the universe does not need constant divine guidance but instead, as some argued, it was like a watch. In this case, once the universe is built and wound up, it runs of its own accord. This changed who God is – no longer a father figure, but the Great Watchmaker. God or some force initially created the universe, thereafter leaving it to operate according to laws of physics that could be discovered by humans.

In addition to Copernicus, the rise of modern science was ushered in by numerous other figures. Francis Bacon (1561–1626) was an English philosopher and scientist as well as an important political official under King James I. In his book, *Novum Organum* (1620), he asks how we can obtain better certainty in knowledge. He discusses how "idols" or biases of the mind can cloud and obscure clear thinking. He lists four idols – of the tribe, the cave, the marketplace, and the theater – as distorting thinking and reason.

The idols of the tribe are natural distortions arising from human nature. These are the tricks the senses can play, mistakes that result from the fact that human senses are like a distorted mirror. One is unable to perfectly grasp or understand the world because one is often tricked by experiences of the senses. The idols of the cave are the biases or distortions found within each individual. Each person has a unique perspective or limits, which lead to misperception of the world in various ways.

The third idols are those of the marketplace. These are distortions that arise as a result of human interaction.

Finally, there are the idols of the theater. These are the distortions that arise from biases in philosophical and religious dogma. Since each of these idols contributes to limits on human understanding, Bacon advocates new foundations and methods of inquiry to address them. New methods of inquiry, based on science and reason, can liberate humans from the four idols. Bacon's arguments thus gave rise to the scientific method.

One specific idol Bacon attacks is the assumption that the universe has a final purposive cause. If one can explain motion and phenomena without assuming a God – as suggested by Occam's razor – then do so.

Bacon rejected appeals to final causes or telos for the universe as merely one example of a religious idol of the theater. The universe is not purposive in the way Christians think. It is completely without final causes, containing only mechanistic ones, which can be discovered, but not through the Scriptures or employment of deductive logic. Instead, Bacon argues for use of the senses, guided by scientific and experimental methods, to check the several idols and gain a real understanding of the universe.

The implication of Bacon's arguments is that God is unnecessary to explain nature or man, and that a natural philosophy dispensing with the dogma of theology is better able to explain the universe and develop human understanding. He thus offers the beginnings of the scientific method, a new way to gain and guarantee knowledge.

Bacon's arguments threatened the Church in two ways. First, by criticizing faith and religion, Bacon undercut the basis of the legitimacy used to support the pope's power. That is, one could appeal to reason to question religious authority, including that of the pope's. Second, by refuting the idea of final causes, Bacon destroyed the foundations of Christian cosmology, which supported the pope's claim of authority as coming directly from God. His arguments, in both cases, were even more of a threat to the universalism of the Church than was Protestantism. He not only made religion an enemy but threatened to reverse the relationship between faith and reason that had sustained the Church for 1,600 years.

In addition to Bacon there was René Descartes (1596–1650). Descartes was a French philosopher who continued to develop many of the arguments about science the former had articulated. Like Bacon, Descartes consciously sought to reconstruct or provide a new beginning for science and knowledge. In his first major philosophical work, *The Discourse on Method* (1637), Descartes noted that the human mind, being capable of errors, necessitates a method to check them. He sought to construct more certainty for knowledge, based on specific rules of inquiry.

His better known book was his *Meditations on First Philosophy* (1641). Here, he searches for the basic principles of all knowledge by finding something he cannot doubt. But the senses cannot be trusted because they can deceive. One can think one perceives the world truly but actually be dreaming, or, worse yet, God might be a great deceiver, or "malignant genius," tricking humans constantly. Descartes continues to doubt all that can be doubted until he finds something that cannot be questioned, even if God is indeed an evil genius. That "something" is the processing of doubt and existence itself. He contends that he cannot doubt his doubting and that if he asserts that, then he exists. In Descartes' language, he states: "*Cogito, ergo, sum*" – "I think, therefore I am." Descartes' conclusion was significant in that the thinking self became the starting point for reconstructing human knowledge and foundations. It is an appeal to reason and rationalism, located in the mind, to substantiate the world. This viewpoint is in clear contrast to the Christians, who had started with God as the basis of knowledge. Descartes reverses the Christian order altogether – it begins with humans.

Finally, there was Isaac Newton (1642–1727), an English mathematician and physicist whose work altered seventeenth-century understanding of the world. His *Principia Mathematica*, or *Mathematics Principles of Natural Philosophy* (1687), reformulated explanations of how bodies in the universe interacted. Previous to Newton, gravity had been explained as a universal attraction of all objects to fall toward the center of the universe, that is, the center of Earth. But when Copernicus displaced Earth from the center, that explanation no longer sufficed.

Newton's three laws of motion helped provide an explanation of gravitational forces. These forces, for Newton, are part of nature itself.

Newton's universe, described in terms of physical forces and mathematical equations, however, means that religious final causes are not crucial to an understanding of everyday occurrences. Although Newton may still have seen God as the ultimate creator of the universe, his God and universe are much different from those of earlier Christians. God is an architect, a mathematician, or a great watchmaker, and the universe is like a great watch. Once God created or wound up the great watch of the universe, it started running on its own, according to laws, rules, and properties that man can ultimately understand and describe.

The implications of Newton's claims were that the natural laws of the universe are not moral but mathematical. One could explain nature not by reference to God as the final cause, but through connections among force, mass, and acceleration.

The Study of Andromeda in a Pre-scientific and Pre-telescopic Era

The mythic and pre-scientific character of much of astronomy until the sixteenth and seventeenth centuries is one of two factors setting the context for our exploration of the Andromeda Galaxy. The other characteristic is that until 1610, when the telescope was invented, observation of the skies was done with the naked eye. Both of these factors are critical to interpretations of what Andromeda was and its role in the universe. The mythic pre-scientific observation of the sky meant that the constellations and other celestial phenomena were cast, as described earlier in this chapter, as gods or other types of living entities that helped to explain change and human affairs. There was an intimate connection between the sky and Earth and humans saw themselves very much connected to what they saw over their heads. Given the lack of light pollution and the prominence and visibility of the sky around them, it is not a surprise that ancient civilizations saw themselves as so directly impacted by and connected to the stars.

Second, until Galileo and the invention of the telescope, observation of the sky was essentially done with the naked eye. This does not mean that there were not some types of astronomical instruments to aid in observation. For example, some see the ancient Egyptian pyramids and the Incan temples and observatories of Mexico and Central America as astronomical tools. Similarly, the Babylonians had their own observatories. Some also speculate that Stonehenge, a circular construction of large stone slabs built on the Salisbury Plain in England beginning in 3000 B.C., had astronomical uses in addition to serving as a burial place. Evidence for that is the Sun shines through the rocks at certain angles during solstices and equinoxes. There is also some evidence that Stonehenge could be used to predict eclipses and other astronomical events. However, whether its use was primarily designed

FIGURE 2.4 Stonehenge.

36 The Andromeda Galaxy and the Rise of Modern Astronomy

FIGURE 2.5 Astrolabe.

for astronomical purposes is debatable, especially since so little is known about who built it and for what purposes.

In addition to structures such as Stonehenge and the pyramids, prior to the telescope there were many other instruments designed for astronomical purposes. The most simple and oldest may be the sundial, which can be found in cultures around the world. The ancient Egyptians used the merkhet to keep and record time. It did that by tracking stars in the sky. In Mesopotamia unearthed clay tablets reveal astronomical tables and numbers that described the location of the stars and planets. These tablets were both chronicles of events and used as predictive tools for time and astrological purposes. Dating from the ancient Greeks astrolabes were used. Astrolabes created models of the skies.

Astrolabes generated two-dimensional representations of the sky on paper, and they had parts that could be manipulated to simulate the movement of the stars and planets in the sky. By Ptolemy's time they were quite sophisticated and were regularly used as astronomical tools. Perhaps because astrolabes employed many circles to represent the sky Ptolemy and others increasingly employed epicycles to explain planetary movements in the sky.

Prior to the telescope there were additional tools also used to observe the sky. Sextants and quadrants were fashioned by sailors to note latitude and determine location. Tycho Brahe designed increasingly more accurate sextants and quadrants to use in his sixteenth-century Danish observatories at Uraniborg and then Stjerneborg. Overall, a host of observational aids were employed for astronomical purposes, but all of them essentially employed naked eye and unmagnified observation of the sky. One could not get a magnified look at the Moon to see what its craters and "seas" looked like. Jupiter was a bright star, but no one could see its moons or Saturn's rings, or the phases of Mercury or Venus as they revolved around the Sun. The Milky Way was a haze in the sky, and it was not clear that it was composed of billions of stars. In short, the sky was limited as to what the human eye could see – to about sixth magnitude brightness. Beyond or below sixth magnitude, the universe did not exist, and even within it, what was seen was limited to what a good eye could observe on a dark night. Given all this, what are the early accounts or records of Andromeda Galaxy?

As an approximately 4.0 magnitude fuzzy patch in the northern skies, M31 must have been visible to many ancient cultures. Without the light pollution of today it would have shown even more clearly thousands of years ago. Yet records documenting its existence are scant. With the attention that the Egyptians, Chinese, Native Americans, and other cultures placed on observing the sky it would seem probable that one or more of these cultures would have seen Andromeda in the sky and would have recorded or noted its existence. One would think it would have been attached to some mythology about the sky or have it included it in some constellation or sky chart. Yet current research of the astronomy of these ancient cultures produces little evidence or discussion of the Andromeda Galaxy.

North American Native Americans located in the Great Basin area that includes mostly Nevada, Utah, Oregon, and parts of Arizona and perhaps New Mexico created petroglyphs. These are stone carvings, etchings, or art. Petroglyphs have also been found in other parts of the United States, including in Minnesota, and all of the date back hundreds or thousands of years. Some scholars

FIGURE 2.6 North American Indian petroglyph.

contend that these petroglyphs are astronomical tools or otherwise record the sky. Among the Great Basin petroglyphs there appears to be some that scholars contend depict the area of the sky that includes the present constellations of Perseus and Cassiopeia, both situated near the Andromeda Galaxy. Other examples of Native American rock art depict the part of the sky that include the region of the Andromeda Galaxy, but the galaxy does not appear to be identified as an object. The Navajo constellation "Thunder" incorporates the area that includes Pegasus and Leo, again suggesting familiarity with the Andromeda region of the sky. Again, no clear reference to M31.

The Chinese also were fascinated by the sky. Solar eclipses were considered sacred predictions for the fate of emperors, and thus the Chinese sought early on to predict them. At least as far back as 206 B.C. they could predict solar eclipses. Some evidence suggests they were able to make eclipse predictions even centuries earlier than that.

As far back as 4000 B.C. the Chinese were producing maps of the sky. They divided the sky into 4 quadrants and then each was divided into 7 parts, producing 28 so-called mansions. The quads were given animal names and the mansions were used to chart the movement of the Moon across the sky. One of the mansions, a tiger, was associated with the autumn, and it included one mansion, 奎, Kui, which includes part of the Andromeda constellation. The Chinese also identified other mansions in the area, and it would seem that they would have included the Andromeda Galaxy. But again, there is no pictorial reference to it. Similarly, the ancient Hindus also charted the sky and the zodiac, and they produced constellations in regions very close to Andromeda, and the same was true for the Egyptians. In fact, it appears all of the ancient cultures far back into pre-history produced star maps and charts used for predictive purposes, mostly to forecast time and the seasons, but none of these maps conclusively labels the Andromeda Galaxy.

Yet there is one possible reference, and that is from ancient Babylonia approximately 2500–3000 B.C. In one of the ancient tablets there is reference to a constellation referred to as a stag or horse. This reference seems similar in identification to Pegasus. The tablet then references "dusty stars which stand in the breast of the horse." It is possible the text is identifying the Andromeda Galaxy, but this is only conjecture or a hypothesis. Again, there is no conclusive proof this is what is being identified. However, given the detail and attention given to the sky in Mesopotamia, it is not inconceivable that this is one of the first recorded instances of the Andromeda Galaxy. The description and sky location seem correct in terms of identifying it as the object in question.

As noted in the introduction to this book, the ancient Greeks explained the constellations, including Andromeda, in mythic terms, seeing them as the byproduct of a struggle among the gods. However, there are no extant depictions of the constellation or M31 from the Greeks. Some speculate that the Roman poet Festus Avienus in the fourth century B.C. might have identified it when he referred to a chained constellation and talked about "thin clouds tie her arms with twisted knots."

While M31 is a naked-eye object in the sky, the first recorded instance of the galaxy appeared in Al-Sufi's *Constellations of Fixed*

FIGURE 2.7 Al Sufi's representation of Andromeda (fish nostrils represent M31).

Stars in 964 B.C. M31 is the nostrils of the fish. Al-Sufi called Andromeda a "little cloud" in his *The Book of Fixed Stars*.

A subsequent version of the constellation appeared in a Latin version of Ptolemy's *Almagest*. Here, a woman and a fish are again used to depict Andromeda. However, here M31 is not fish nostrils but a little smudge to the right (in the figure below) of the nose. This smudge was described through the European Middle Ages as a "nebulous spot." A Dutch sky map from circa 1500 also references Andromeda, but beyond noting its existence, little else is said about it. Giovanni Batista Hodierna, court astronomer for the duke of Montechiaro, is reputed to have rediscovered Andromeda and noted it in his 1654 *De Admirandis Coeli Characteribus*. In *De Admirandis* he classifies nebula into three groups – Luminosae, Nebulosae, or Occultae – with Andromeda situated in the third category.

Thus, prior to the invention of the telescope, very few clear and direct references are given to Andromeda. Perhaps it was not seen, but this is unlikely. More likely this faint object did not fit into the some mythological explanation for the sky and therefore it was ignored.

FIGURE 2.8 Andromeda as depicted in a Latinized version of Ptolemy's *Almagest* (Andromeda is located to the right of the fish nostrils).

Early Modern Depictions of Andromeda

It would not be an exaggeration to state that the telescope revolutionized astronomy, science, and how humans understood their role in the universe. Without the telescope the sky was limited to what the human eye could see. It was limited to a closed world of the sixth magnitude, a world of five planets (besides Earth), one Moon, and a belief that our world was the center of the cosmos. The sky that astronomers in 1600 saw was essentially the same one that the ancients viewed, including Ptolemy. It was thus a geocentric cosmos where Earth stood still, and the Sun, the planets, and the stars all revolved around in circles and epicycles. But the invention of the telescope turned this closed world into an infinite universe, one that confirmed many of the assumptions of Copernicus and his view that Earth was not at the center.

There is some dispute over who built the first telescope. Hans Lippershey (1570–1619), a Dutch lens crafter, in 1608 is often credited as the first person who placed two lenses together to create the first telescope. Others credited or claiming to have invented the telescope include Zacharias Janssen and Jacob Metius (1571–1628), both also of the Netherlands. Lippershey's first telescope placed together convex and concave lenses to produce a simple refractor telescope. This simple telescope yielded a magnification of no more than three times. Using it, you could see three times as far away as the plain human eye could. Lippershey in 1698 applied for a patent for his instrument with the States-General of the Netherlands, asserting that it was "for seeing things far away as if they were nearby." His patent was approved several weeks before Metius's was.

Originally the telescope was viewed as an important and promising military tool. It would allow for the spotting of troops or ships earlier than would be possible with the naked eye. But the astronomical value of the telescope fell to Galileo to discover.

News about the telescope spread quickly from the Netherlands, across Europe, and to Padua, where Galileo was. Galileo first heard news of the telescope in 1609. He placed two lenses into a long tube to produce a simple refracting telescope. His first telescope was not much more powerful that Lippershey's – it had 3× magnification. His second telescope yielded a 10× magnification. When he demonstrated it to local officials, Galileo's salary was doubled, and he became Padua Chair (for life) of Mathematics. The third telescope had a magnification of 20×, and it was constructed out of a strips of wood glued together. It was perhaps this third telescope that was used for many of Galileo's discoveries.

Even though Galileo did not invent the telescope, he did improve upon the original design. This is an accomplishment in itself. But the bigger achievement for Galileo was what he did with the telescope. He turned it to the sky for observation. It was this act that forever changed astronomy. During 1609–1610 Galileo turned to many objects in the sky. He observed the Sun and saw spots on it (and in the process nearly blinded himself). He saw the craters and mountains of the Moon, too. But more significantly, he observed the planets. He saw the phases of Venus and the rings of Saturn (although he did not realize they were rings). Additionally, from his sketchbooks and notes he might have been the first to observe Neptune, although he did not realize what he was seeing at the time and instead thought it was just another star. But most importantly, he observed Jupiter.

It was not the observation of this gas giant that was so significant; instead it was his discovery of the planet's four main moons circling it: Callisto, Europa, Ganymede, and Io. Seeing these moons circle Jupiter provided to him evidence that Copernicus was perhaps correct in that Earth was not at the center of the universe and that it, along with the other planets, orbited the Sun.

Galileo reported many of his findings in his 1610 *Sidereus Nuncius* or *Starry Messenger*. The book also reported on observing many star clusters, including Pleiades, as well as the Milky Way. The book's long term impact was enormous for Galileo. It led to his eventual censure by the Catholic Church, of course. For astronomy, it expanded, enlarged, and added to the view of the universe and provided evidence that Copernicus was correct in that the universe was not geocentric and therefore humans were not at the central point in the cosmos. Galileo's observations added evidence that the cosmos of Ptolemy and of the Catholic Church was not correct, therefore shaking the foundations of the Great Chain of Being that held together the political and theological order of the universe.

There is no evidence or any reports that Galileo observed the Andromeda Galaxy with a telescope. However, the first known telescopic view of it occurred in 1612. On December 15, Simon Marius described it as looking like a "candle shining through horn," referencing in that description the common practice then of using a translucent horn to diffuse light in lanterns. He also stated Andromeda looked "like a cloud consisting of three rays; whitish, irregular and faint; brighter toward the center." Given that technique of illumination, Marius's description of M31 was quite descriptive and accurate.

The next major figure to have noted the appearance of Andromeda was Edmund Halley in 1716. Halley (1656–1742) is best known for predicting the orbital term of the comet – Halley's Comet – named after him. He described Andromeda by stating it was "nothing else but the light coming from an extraordinary great space in the ether, through which a light medium is diffused that shines with its own proper luster."

In 1749, Guillaume Le Gentil (1725–1792), a French astronomer, was another major figure to observe Andromeda. In doing so, he discovered what eventually would be M32, a companion galaxy to Andromeda. But while in search of comets Charles Messier produced a list of objects he wished not to confuse with them. Of a list in excess of 100 objects, Andromeda was number 31, making it to this day still known as M31. Messier described his 31 object

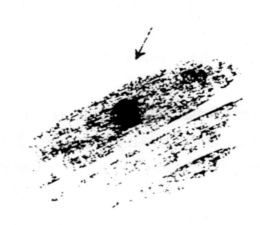

FIGURE 2.9 Lord Rosse's drawing of M31 (about 1871).

as looking like "two luminous cones or pyramids opposite at their bases." Messier also sketched M31, revealing two companion galaxies with it that would eventually be labeled M32 and NGC 205.

After Messier, William Herschel observed M31. He noted of it that: "[T]he stars which are scattered over it appear to be behind it, and seem to lose part of their luster in the passage of light through the nebulosity."

Finally, in 1871, Lord Rosse (William Parsons) produced one of the first hand drawings of M31 that seemed to show the central region of the object. He produced this figure through observations in his 6-foot reflector telescope.

Thus, by the nineteenth century, the shape of Andromeda had emerged. Telescopic observation yielded a spiral, with a brighter spot in the middle. Andromeda had companion objects next to it, and all three were distinguished as separate celestial entities. The view of Andromeda had clearly changed from the days of the ancient mythic naked-eye observation. It was natural to turn the telescope to examine this fuzzy object in the sky, but what it was remained a matter for debate. Determining what it was in the late nineteenth and then in the twentieth century would be a major enterprise for astronomy, changing the field and our concept of the universe.

3. A Single Closed Theory of the Universe

When we speak of the universe, what do we mean? Today it is thought of as an entity of nearly infinite size and dimensions, approximately 13.7 billions of years old, full of perhaps trillions of stars organized into a countless number of galaxies. It is a universe composed of many parts. There is no center, no central point, and gazing in any direction from any point one would never see the edge.

Yet that is not how the universe was conceived or thought of up to and through the nineteenth century. From ancient times through the early part of the nineteenth century the universe was much more structured and ordered in a hierarchal fashion. Its logic or structure in Christian Europe was the Great Chain of Being described in the previous chapters. It was a universe with Earth as a fixed center, with all of the rest of the cosmos orbiting around it. Earth was also located between heaven and hell, standing motionless in a finite universe. The universe was truly a singular entity – with fixed dimensions, with an age defined by creation through God as defined by the Genesis story. There was one human race, located on one Earth, located at the center of one Solar System, in one galaxy, in one universe. The universe was a closed and finite body. More or less, this was the image of the universe that began with the ancient Greeks and carried forth until the nineteenth century.

Thus, well into the nineteenth century, even after the invention of the telescope, astronomy was dominated by several assumptions that influenced knowledge and depictions of Andromeda. These assumptions were: (1) belief that the universe's distances and age were much shorter than they are now; (2) that the Milky Way was the only galaxy in the universe; and (3) that nebulae were merely patches of unresolved stars connected to our galaxy.

Sizing Up the Universe

Perhaps the most basic question of cosmology, metaphysics, or theology is one that asks what is the universe and why does it exist? For philosophers, this question is part of the branch of their field called ontology, which is the study of being or existence. For these individuals, the most basic question to ask is simply "Why is there something as opposed to nothing?" This question asks about why life and the universe exists, as opposed to there being nothing. But in seeking to understand how the universe was conceived or depicted by nineteenth-century astronomers, yet again it is necessary to return to the ancients. It is from the ancient Greeks, Christians, and perhaps even other cultures that the view of the universe of the nineteenth century emerged.

The beginnings of seeking to understand the universe in terms of its dimensions perhaps again starts with the Greeks. As they sought to explain the origins of the universe they also wanted to describe its size. But efforts to do that also begin with a more simple set of questions. Many wanted to know about distances on Earth, how big Earth was, how far away the Moon was from Earth, and also how far away the Sun and the stars were. They were thus interested in basic geographic questions that remain of concern today. They also sought to define distances to celestial objects, which began a quest that persists in astronomy to this day. Astronomers continue to search for better and more accurate tools to measure distances to objects in the universe, employing tools of geometry, physics, and motion, all techniques that have their roots in the ancient Greeks.

When the Greeks thought of the universe, how did they describe it? As pointed out in earlier chapters, philosophers and scientists such as Anaximander and Anaximenes thought the universe was composed of simple or basic elements such as earth, air, wind and fire. Others such as Democritus saw the universe composed of elementary particles labeled by Democritus as atoms. The Pythagorians thought numbers were the reality of the universe, and others, such as Plato, saw something called forms as the reality at the heart of the universe we see. All of them thus believed that there was a core set of fundamental particles or building blocks out of which the universe was composed. The core problem for the Greeks resided in explaining two issues: the one versus the many, and second, the problem of change. Both problems were connected.

The problem of the one and the many simply asked if the universe and the world were one single entity or composed of multiple parts. Was the universe made up of one element, such as earth or air, or were there multiple elements? Moreover, the one and the many asked if the space of the universe was unified, was everything connected, or was there a void or emptiness between objects that comprised the universe. In other words, is there empty space between the sky and Earth? What comes between Earth and the stars, if anything? Is there something or nothing?

The problem or paradox here with the one and the many is twofold. First, if all is one, if everything is in fact connected, then how do we explain how things are different or individual? How do we account for differences we see? Are all individuals, stars, and different points connected? If there is difference or a many, how do we account for it? Why are entities different, and how do they interact? This is leads to the second problem, change.

How can we account for change in the universe? Is the universe static or dynamic? Is there any growth, movement, or motion in the universe?

Depending on whether one approached the universe from the one or many perspective, different answers and problem arose. If, for example, the universe was seen as a one, then change could be explained in simple ways. If everything is connected, then as one entity or part of the universe changed, so would the rest. Motion in one part of the universe would lead to motion elsewhere. It would be like throwing a rock into the water and seeing the ripple effects fan out. Or it would be similar to striking one billiard ball and it hits another and then another. This is simple Newtonian physics that describes how a force applied to an object puts it into motion until another force stops it. Change is simple to account for in a universe composed of one. Since everything is connected, everything moves together. Thus, motions and events on Earth affect the sky and vice versa.

If everything is connected, then astrology makes sense. The motion and location of the stars in the heavens do have an impact or a causal relationship to everything else, including our lives. If the universe is a one, everything seems causally connected. Everything is determined, and perhaps there is no concept or room for free will or choice. Our lives are set and determined by the fate of the universe. In fact, for the Greeks, the Fates were goddesses – Clotho, who spun out our lives, Lachesis who determined the length of our lives, and Atropos, who cut the thread of life and determined our deaths. The three Fates thus determined our destinies.

Yet if the universe was not a one but a many, one could account for free will, choice, and individualism. Our lives were not determined or controlled by the cosmos but by the decisions we made. But a universe of many caused a problem in explaining change. First, as Heraclitus pointed out, if everything was in constant change, there was no permanence. If no permanence, how does one describe reality? The Greeks assumed that reality had to be something – static, unchanging, and fixed. Underneath or behind the world of change there had to be something permanent. If everything always changed, there really was nothing that could be defined as existing. If rivers constantly changed, is a river ever the same river if you step in it the second time? It is different and constantly evolving. Additionally, if the universe is not one, but constantly changing, how does change occur? If there is a separation between Earth and heavens, what connects them? If a void exists, then how can two things separated ever affect one another? How do we explain, perhaps, how or why Earth and the Moon affect one another if in fact a void separates them? The same is true for the Sun and Earth and Earth and the rest of the universe. If everything is separated, then explaining causal connections and interrelationships is impossible. It becomes hard to explain such simple things such as the rising and setting of the Sun, the movement of the Moon and stars across the sky, and even the change of seasons. All imply change, but if there is no connection, no oneness but a many in the universe, then explaining movement and connections become difficult.

The Greeks were stymied by the one and many and the change problem and the problem of distinguishing appearance from reality. All three of these issues persist as scientific and perhaps religious and philosophical problems even to this day. Religion seeks to solve the problem by a belief in a deity who is invoked to explain creation. Yet the "one and the many problem" remains. If God created the universe, how do we account for evil? Can we account for free will in a universe created by God, or are all the choices we make determined by creation? If so, then is anyone really responsible for the sins they commit? These are difficult theological questions. Modern physics and cosmology, too, still seek to explain the origin of the universe, seeing is as coming into being in a single event, the Big Bang, and we still strive to find a single particle or force that accounts for the origins and for the change we see around us. Astronomy thus is still driven by explaining problems that date back thousands of years.

The Greeks initially confronted this problem in their conceptualization of the universe as something infinite. Both Anaximander and Anaximenes described the universe as infinite and one, despite the fact that they saw it composed of constituent parts such as earth, air, wind, and fire. The infinite cosmos was not composed of any of these discrete elements but instead of all of them. Explaining how out of one the many emerged, including eventually Earth, the moon, the sky, and the stars, was part of what was discussed earlier. These explanations invoked both mythic and anthropomorphic qualities, as well as vortexes, such as heating and cooling. But despite this view of the cosmos as infinite, both of these individuals had to account for distance and size. Both, in other words, wanted to measure how big the infinite was and explain the relative order and placement of entities in it.

Anaximander explains the universe's distance and placement of objects by suggesting that relative weights and temperatures determine location. Earth is located close to us because of its weight. Around Earth there is then water, not as heavy, followed by air, even lighter, and then the heavenly bodies circle, their placement determined by their heat. The most distant entity is the Sun, since it is closest to the ring of fire around Earth. After that, the Moon is closer and then there are the planets and the stars. Anaximander believed the Moon to be located in a ring or orbit 18 times as large as Earth's diameter, the stars in a ring 9 times as large as the diameter of Earth, and the Sun 27 times as large. Additionally, Anaximander declared the diameter of Earth to be three times its depth. Given all of these multiples computed in a ratio of threes, it is less likely his distances were based on real calculations than on some basic assumptions about the order of the universe.

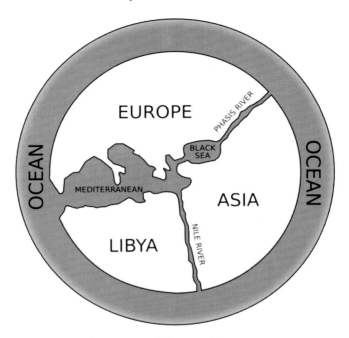

FIGURE 3.1 Anaximander's map of the world.

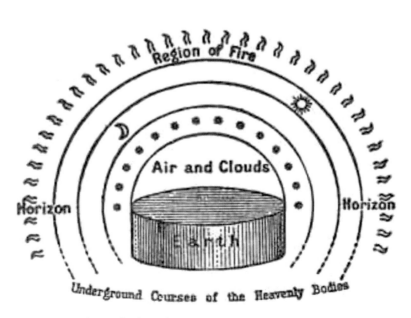

FIGURE 3.2 Anaximander's universe.

Although Anaximander's statements about distances in the universe were based on some preconceived notions of cosmic order and structure, other ancient Greeks did seek to determine distances. Their concern focused on the practical distances between objects on Earth and the size of Earth.

Aristarchus (310–230 B.C.) is most famous among ancient astronomers for pre-dating Copernicus in arguing that the Sun and not Earth is at the center of the universe. But he is also famous for one of the earliest attempts at measuring the distance of Earth to the Sun. He did that employing geometry. Aristarchus assumed that the Sun, Moon, and Earth formed a right triangle during the first and last quarters of the Moon's phases. The angle of Sun–Moon–Earth was assumed to be 90°, and then he estimated the Moon–Sun-angle to be 3°. Using simple geometry and ratios of angles to sides he computed the distance to the Sun to be 20 times that of the distance of Earth to the Moon. His numbers were way off. Today we know the Sun to be approximately 390 times the distance to the Moon. He also calculated Earth's diameter to be 3 times that of the Moon, and that the Sun was 20 times the size of the Moon and 20 times the distance from Earth. Again, these numbers were off significantly.

Eratosthenes (276–195 B.C.) was in charge of the famous library in Alexandria, Egypt, and he too sought to calculate distances by bringing together geometry and geography. His calculations were more accurate than Aristarchus's. Eratosthenes observed that when the Sun was directly overhead at Syene, Egypt, it was 7° south of the zenith in Alexandria. These 7° are 1/50 of a circle, and therefore he determined that the two cities were about 1/50 of Earth's circumstance. Egyptians employed the term "stades" to measure distances, and by common consensus the distance between Alexandria and Syene was 5,000 stades. Doing the math, we get 50 × 5,000 stades = 250,000 stades. By today's standards and knowing the distance between the two cities in kilometers, his estimate of the circumference of Earth was 42,000 km (about 29,000 miles), not far off from contemporary calculations that place it at 40,000 km (25,000 miles).

Other Greeks also sought to calculate the size of Earth, the distances to objects, or the speed of heavenly motions. Both Apollonius of Perga (262–190 B.C.) and Hipparchus (circa 190–120 B.C.)

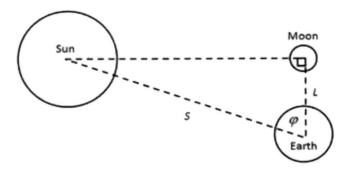

FIGURE 3.3 Aristarchus's measurements of earth.

attempted to calculate and model the motion of the Moon and the Sun. Apollonius in his models assumed some objects such as planets moved along a deferent, or major circle, and then along a secondary one referred to as an epicycle. These two circles helped predict the orbits of planets, and this basic model, along with a geocentric assumption about the universe, were perfected by increased precision and observation through Ptolemy and up to Tycho Brahe and Johannes Kepler before eventually the heliocentric model proposed by Nicholas Copernicus replaced it, rendering the complex patterns of multiple deferents and epicycles unnecessary. Hipparchus modeled the motion of the Sun. He noted that the length of seasons was not the same and that the Sun seemed to move across the sky at different speeds. For example, it took the Sun 94.5 days to move 90° from the spring equinox to the summer solstice but only 92.5 days from the latter to the fall equinox. To account for this Hipparchus hypothesized that Earth was not exactly at the center of a circle around which the Sun revolved. Earth instead was slightly off center, producing an eccentric path for the Sun.

Ptolemy adopted many of these assumptions and models about distance in his astronomical writings. He thus adopted the concepts of deferents and epicycles in order to get an accurate picture of motion for the planets. But like Hipparchus he had a problem in assuming Earth to be located at the center of a group of circles around which celestial objects moved. However, instead of assuming eccentric paths, he invented another idea, the equant point. The equant point was opposite the center of the circle and

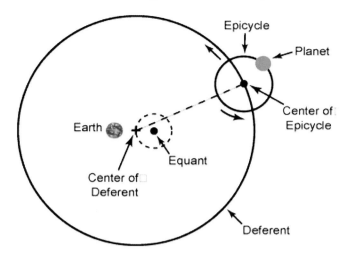

FIGURE 3.4 Ptolemaic depiction of planetary movement.

the location of Earth. The equant point permitted one to account for the different speeds objects moved across the sky. What it assumed is that Earth was not exactly at the center of the circle but off slightly, equally distant from the center as was the equant point. But planetary and other motions, when observed from the equant point, were able to account for the slight variations and motions across the sky and around Earth.

Thus, heavenly motion was still in perfect circles, along with the circles of the epicycles. The model still allowed for a geocentric universe, even if Earth was not exactly at the dead center. Ptolemy's model, as complex as it was, and also as contrary to the modern laws of physics and Ocham's Razor, remained more or less in place until destroyed by Copernicus and his heliocentric assumptions, and Kepler and his arguments that the planets move in elliptical and not circular orbits. Finally, according to Ptolemy, he calculated the distance of the Sun to Earth to be approximately 1,200 Earth radii, and the distance to the stars to be 20,000 times Earth's radii. The Moon varied between 33 and 64 Earth radii. The size of Ptolemy's Earth was even smaller than that of Eratosthenes, with some measures suggesting it was one-sixth smaller than the latter. Ptolemy's heavenly distances were wrong, postulating a universe far smaller than contemporary astronomy calculates.

The Andromeda Galaxy and the Rise of Modern Astronomy

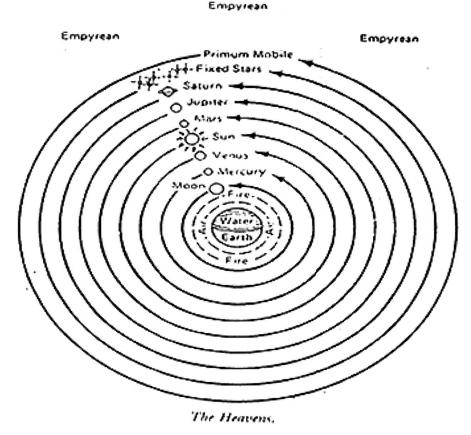

FIGURE 3.5 Ptolemy's universe.

Ptolemy's geographic and astronomical assumptions carried over into the Christian Middle Ages.

This second-century map of the universe looks very much like images similarly created in the Middle Ages.

The model of the universe joining Ptolemy and Christianity is wonderfully depicted by a medieval sixteenth-century woodcut showing the Biblical reference to Jacob. Here Jacob is seen poking his head through the stars and universe, reaching the edge of universe and into heaven. The universe, while vast, was still finite and limited in size. It was not the infinite cosmos of the Greeks, but instead a finite, closed, and hierarchal universe.

A Single Closed Theory of the Universe 55

FIGURE 3.6 Medieval woodcut of Jacob poking his head through the universe.

Efforts to calculate the size of the universe continued through the Middle Ages, Renaissance, and into the nineteenth century. Into the fifteenth century the cosmology and expanse of the universe was still thought of in Ptolemaic terms. It a finite universe of fixed dimensions with the Sun, Moon, planets, and stars all located in fixed places. Yet beginning with Nicholas of Cusa (1401–1464), a Catholic Church cardinal, this cosmology was questioned. Nicholas of Cusa began questioning the span of the universe as finite, arguing not that it was infinite but instead indeterminate. Following him, Copernicus suggested the universe to be immeasurable but not infinite, but even then the expanse of the universe to the most distant stars was estimated in the sixteenth century to be about 125 million miles, or about 20,000 times Earth's radius.

Subsequent to the Copernican reconfiguration of the universe as heliocentric, others began to argue for a more expansive or

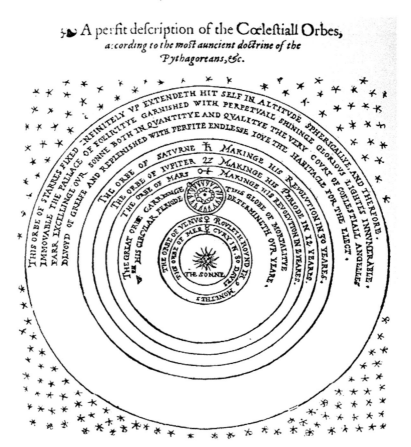

FIGURE 3.7 Thomas Digges' map of the universe.

infinite cosmos. Giordano Bruno (1548–1600), an Italian Dominican friar, first suggested that the universe might actually be infinite. Additionally, Rene Descartes, a French philosopher and mathematician noted for inventing one type of calculus, described the universe as infinite and uniform.

After Copernicus, the first individual who constructed a chart of a heliocentric universe was Thomas Digges (1546–1595), a British mathematician. Digges's diagram seemed to postulate a Copernican universe more expansive and open than the closed one proposed by Ptolemy. Digges describes a universe where the stars are infinitely far away from Earth.

FIGURE 3.8 M12, globular cluster in Hercules.

Although his diagram of the cosmos represented an improvement over previous ones, there were problems. First, his model seemed more based on theological than astronomical assumptions in that it still followed the basic order of the Christian universe with all the stars of equal fixed distance from the center of the universe. The stars thus still seemed located closest to heaven and God. Second, because Digges' model presupposed stars as equidistant from Earth, Digges ran into another problem. Specifically, if there are stars all around Earth and they are constantly shining, why was there darkness at night? Should not all the light from all these stars reach Earth at the same time, thereby keeping the sky alight even at night? If one assumes stars at an infinite but still fixed and equidistant point from Earth then this is definitely a problem. Digges thus may have been the first astronomer to address what has become known as Olber's paradox. The paradox is named after Heinrich Olbers (1758–1840), a German astronomer, who raised this question about the night sky in the nineteenth century. He pondered why is the sky dark at night if it is full of stars all emitting light? Addressing his paradox would

FIGURE 3.9 A 1890 photograph of M31 by Isaac Roberts.

eventually perplex nineteenth-century astronomy, forcing it to reconceptualize distance and the origins of the universe.

Despite claims about a finite or infinite universe, some efforts to measure distance were employed. Recall that Aristarchus tried to measure the distance from Earth to the Moon. He did that using geometry and a technique called parallax. The concept of parallax is simple. Hold out a finger at arm's length and then blink or close and open one eye at a time. The finger appears to move. What has happened is that the distance between the two eyes and the finger form three parts of a triangle. If one knows the distance between the two eyes one can eventually compute the angles that form the triangle, and then using simple trigonometry eventually the distance to the finger can be calculated. The finger appears to move quite a bit when blinking the eye simply because the distance is small and the apparent angle formed by the eyes and the finger are significant. Aristarchus used this parallax method to compute distances and the size of the Moon and Sun, although he got both wrong.

Another attempt to compute stellar distance and therefore potentially the size of the universe was made by Tycho Brahe.

On November 11, 1572, a bright supernova appeared in the constellation of Cassiopeia. Brahe assumed the object could not be a star because of the Christian and Ptolemaic assumption that the cosmos, or at least the stars, were permanent and fixed. Thus the object must be something else, and since it was not a star, it must be rather close to Earth. To estimate the distance Brahe used parallax. He did that by observing the supernova 12 hours apart. Because of Earth's rotation, or for Brahe, the rotation of the heavens, since he still believed in a fixed Earth with the heavens rotating, the supernova should change position in the sky relative to the position of other stars.

Despite his careful measurements, he was unable to detect much parallax and therefore compute the distance. He tried again in 1577 with a comet, but again to no avail. The problem was not with the parallax method, per se. Instead the problem was that the objects were much farther away than Brahe thought, and the utility of parallax to measure the distance to such objects is limited. The reason for this is that for very distant objects their apparent shift in the sky is so miniscule. Computing distances in the universe is a problem for astronomers, and new techniques beyond parallax have been developed that are able to address some of the larger distances. One of those techniques, involving Cepheid star variables, was discovered by Henrietta Leavitt (1868–1921), and her technique would be critical to establishing the distance to M31 Andromeda and establishing it as a distinct galaxy in the universe.

However, prior to that measurement in the twentieth century, other astronomers sought to estimate the size of the universe, or at least the Solar System, coming to varying answers. Christiaan Huygens (1629–1695) estimated the distance to Sirius to be that of 27,664 times the distance of the Sun to Earth. He made this estimate by placing a screen between him and the Sun and reducing the size of the Sun such that it would only be as bright as Sirius. But the brightness of the Sun made this technique dangerous (NEVER observe the Sun with your naked eye or through a telescope without appropriate protections) and inexact. James Gregory (1638–1675), a Scottish mathematician, also compared Sirius to the Sun, concluding that the former was about 83,000 times as far from Earth as the Sun was.

In 1671 Pierre Cassini estimated the distance of Earth to the Sun to be 87 million miles. This was 6 million short of the approximate 93 million now calculated. But given that Ptolemy estimated the Earth–Sun distance to be 4 million miles, the 87 million was a vast improvement and enlargement over previous estimates. Transits of Venus over the disk of the Sun during 1761 and 1769 were also employed to measure the distance of Earth to the Sun, but estimates varied widely by millions of miles.

Sir William Herschel (1738–1822) was a German-born British astronomer. Along with his sister Caroline (1750–1848), they made many discoveries. These discoveries included the planet Neptune, which William originally wished to call George after the King of England, along with the planet's principal moons Titania and Oberon as well as Saturn's Mimas and Enceladus. William Herschel built many telescopes that he used to search the sky, including a 40-ft-long telescope with nearly a 50-in. mirror. These were the most powerful telescopes of their time. But in addition to his discovery of Neptune he also undertook extensive studies of Mars, and he examined many deep space objects. He had concluded that our Solar System was moving and speculated that the Milky Way was a huge disk of stars. Using a variety of telescopes, he is credited with discovering and cataloging approximately 2,400 deep space objects. All of these objects were referred to as nebulae at that time, and it was only later in the twentieth century that distinctions among objects, such as between nebulae and galaxies, were made.

Herschel was one of the first to undertake extensive study of Charles Messier's objects, including that of M31. In observing the Andromeda Nebula he initially agreed with claims by others such as Galileo that all nebulae were simply unresolved stars and that with sufficient magnification he could or would be able to see them. Eventually he conceded that perhaps nebulae consisted partially of gas. In reference to Andromeda, he believed that there were unresolved stars in it, but these were covered or layered by gas, too.

More significantly about Andromeda, he offered an estimate of its distance from Earth. According to Herschel, it was no more than 2,000 times the distance to Sirius the star, making it no more than 17,200 light-years away. His distance to Andromeda clearly enlarged the size and distance of the universe, but it falls far short of the 2.54 million light-year estimate astronomers now have.

In addition to Herschel, other astronomers, including Christiaan Huygens (1629–1695), a Dutch mathematician, and German Frierdrich Bessel (1784–1846), experimented with stellar parallax to estimate distances to the stars and to estimate the size of the universe. Their conclusions, as well as those of other astronomers in the nineteenth century, were inexact when applied to relatively close stars, but they also produced weak results to more distant objects. This was a consequence in part of the great distance of these distant stars, yielding almost negligible parallax. However, given the resolving power of telescopes also in use during this time, the most distant resolvable objects were not as distant as those detected today. As a result, through most of the nineteenth century the universe was depicted as a much smaller entity than it is today, with an age assumed to be much shorter than estimates offer now. In fact, into the early twentieth century, debates over whether the universe was finite dominated astronomy journals.

Thus, Herschel's estimate of Andromeda's distance is significant in many ways. Most importantly, what it suggested was that by the nineteenth century the scope of the universe had expanded. It was no longer the closed and finite structure that the Christians and Ptolemaic astronomers had assumed. In the course of nearly two millennia cosmic distances had expanded, such that by 1,800 talk about its size was in light-years and not miles or small ratios to Earth's radius. The universe was vast, and Herschel's Andromeda estimates expanded the size, but it was still considered small by today's estimates.

The Milky Way as an Island-Universe

Astronomers today refer to Andromeda (M31) as a galaxy. Yet that designation is a recent twentieth century label, ascribed to the work of Edwin Hubble in the 1920s. Prior to that, and through the nineteenth century, almost all celestial phenomena beyond the planets, stars, and a few other objects were labeled as nebulae. The same is true with Andromeda. Until the 1920s it was called the Andromeda Nebula. Charles Messier in the cataloging of the objects he saw referred to them as nebulae, regardless of whether they were what we now call star clusters or truly nebulae. William Herschel similarly called many of the objects he saw nebulae. Both of these individuals referred to M31 as a nebula.

The reason for this is that it was assumed that there was only one galaxy in the universe. Thus, a second assumption throughout the nineteenth century and earlier revolved around the belief that the Milky Way was the only galaxy in the universe. In fact the consensus as late as 1900 was that perhaps the Milky Way was the sum of what the universe was. Agnes C. Clerke (1842–1908), a famous and astute historian of nineteenth-century astronomy, contended in her 1890 book *The System of Stars* that no competent astronomer could assert that the nebulae that were observed constituted external galaxies. She contended that the state of scientific progress had clearly shown this claim or belief to be false. There were several reasons supporting Clerke's claims.

The first claim returns back to the idea of the Great Chain of Being and the Christian conception of the universe. This view of the universe originally placed Earth at the center of the universe, and that of course meant this planet was also at the center of the Solar System and the galaxy. The Sun, the Moon, the planets, and the stars were simply considered from the Greeks through Ptolemy and to the Christians to be part of one universe that had only one galaxy in it. In fact, until telescopes revealed the haze in the sky to be stars that constituted the Milky Way Galaxy, no one even talked of or conceptualized that the universe might be composed of a galaxy or galaxies. It was not until the seventeenth and eighteenth centuries, based on the work of William Herschel and others, that the universe was seen as composed of a mass of closely related stars that we have come to call the Milky Way Galaxy. No maps of the universe from the Greeks though Christian Europe depicted more than one galaxy; all of the stars were simply equidistant from Earth in some circle.

To conceive of there being more than one galaxy is as theologically threatening as thinking Earth is not the center of the universe. If Earth is only one of several planets, then that questions the special or unique place that humans have in the cosmos compared to God. Now imagine the repercussions of conceiving of more than one solar system with more planets. Now expand upon that – imagine scores of other galaxies besides the Milky Way! Each galaxy might contain other solar systems with other planets and, perhaps, other humans or living creatures. Such an image raises religious havoc. Genesis talks of God creating the Earth and humans in 7 days. No mention is made of other Earths or humans. The Bible also speaks of humans created in God's image. But if there are other galaxies, how does this square with what the Bible says? To admit the possibility of other worlds raises profound questions for the Great Chain of Being. This possibility remains a problem even today as exoplanets are discovered and talk of life on Mars or other places in the universe is raised. Thus, throughout the nineteenth century, no religious conception of the universe seriously considered other galaxies, and this assumption carried over into astronomy.

A second reason why the astronomical consensus was that there were no other galaxies beyond the Milky Way rested with the state of scientific knowledge at the time regarding the size of the universe. As discussed above, through the eighteenth and into the early nineteenth centuries the estimates of the size of the universe made it quite small compared to contemporary measurements. From Ptolemy to Herschel the size of the universe was consistently expanded from a finite to infinite entity, but still it was not envisioned as being very large. Herschel had argued that the Andromeda Nebula was not much more than 17,000 light-years away. A distance this small hardly suggested a truly infinite universe. While M31 was seen as 2,000 times further from Earth than Sirius was, no nebula were seen in the distances of millions or billions of light-years away. Everything in the sky was relatively close to one another – at least by today's measurements – and therefore everything observed was considered to be part of one galaxy, the Milky Way.

The dominate idea then of nineteenth-century astronomy was that the Milky Way was a single universe. The Milky Way was the sole galaxy within the universe. In fact, one could simply state "Milky Way = universe." The cosmos was no more than all of the phenomena observed or that existed as being part of one galaxy and one universe. The universe, vastly larger than how the ancients depicted it, is small by present-day accounts, and into the nineteenth century there was matter (stars, planets, comets, and other phenomena) and space between them, as constituting one galaxy, one island-universe. The belief in one galaxy in the universe thus dominated astronomical thinking into the nineteenth century, and the best available scientific measurements of the time seemed to confirm this orthodoxy.

This is not to say that some did not question the single universe theory. Philosopher and cosmologist Immanuel Kant and French astronomer Pierre Laplace proposed in the nineteenth century a rival theory of island-universes. For them, the universe was composed of many distinct galaxies. This theory, discussed more fully later in this book, proposed a cosmos of many galaxies. Astronomy and astronomical research in the late nineteenth and early twentieth centuries would increasingly become gripped by this debate, only to be resolved by research conducted by Edwin Hubble in the 1920s when measurements of the distance and size of the universe made it increasingly impossible to contend that the universe was one galaxy. In reaching that conclusion, M31 would be center stage. It would be measurements of the distance to the Andromeda Nebula that would eventually lead to the conclusion it was a distinct galaxy, thereby destroying the single universe theory. Suffice to say then, astronomical observations of the Andromeda Nebula served as bookends at the start of the nineteenth and then the middle of the twentieth century. Herschel's measurements seemed to confirm a single universe theory; Hubble's confirmed the island-universe claim.

Nebulae as Unresolved Stars

Prior to the invention of the telescope the universe was composed of the stars the naked eye could see. With the best eyes seeing stars to about sixth magnitude, the assumption was that there were only a few thousand stars in the entire cosmos. Enter the telescope. When Galileo turned it to the sky his telescopes had limited magnifications of simply a few powers. More stars could be seen – stars the human eye never saw before. With more and more powerful telescopes even more stars could be seen. As the sky was observed by others such as Charles Messier he discovered new objects in the sky. Hoping them to be comets, he detected slightly more than 100 objects that he cataloged. He did this so that he would not confuse them with comets. Some of the objects he saw were visible, although not well with the naked eye, such as the Andromeda Nebula, others only visible with the telescope. But as he observed these objects with limited magnification or light-gathering power, it was not always clear what they were.

For amateur astronomers even today, using a small telescope to observe Andromeda one does not get the wonderful pictures of a swirl or spiral that we see from the Hubble telescope or other large ones at major observatories. Those pictures are the product of major telescopes, countless hours of time-lapsed photography, or even the stacking of many photos upon one another. A small backyard telescope might reveal M31 to be a hazy blob. The same is true of globular cluster such as M13 in the constellation of Hercules.

One of the major frustrations for amateur or backyard astronomers with small telescopes is that they expect to see in their instruments what the major observatories show. Yes, a small telescope reveals the rings of Saturn and the moons of Jupiter, but for deep space observing, it takes more sophisticated instruments to see much more. It is this frustration that probably leads to many telescopes collecting dust in garages and closets!

Yet the views that many backyard astronomers have of the sky today is perhaps superior to that of Galileo and Messier. Their telescopes often had inferior optics compared to today, and their telescopes were often smaller. With smaller lenses for refractors, and smaller mirrors for reflectors, the resolving power of these 'scopes' was limited. Thus, many of the objects observed in the sky were not seen clearly. But as new and more powerful telescopes were fashioned by individuals such as William Herschel, more and more objects could be seen. He cataloged approximately 2,400 nebulae. As he developed better telescopes he could see these nebulae better; yet despite increased magnification and light-gathering power, it was often difficult to ascertain for certain what these objects were. A blob of something in the sky was assumed to be something, but what? The belief was that many if not all of the nebulae were unresolved stars.

A third assumption held by astronomers at the beginning and to the close of the nineteenth century was that nebulae were simply unresolved stars. This assumption had carried over from the days of Messier and Herschel The belief was that if in fact one could obtain a telescope with sufficient light-gathering power one would eventually be able to peer beyond the hazy nebula and see the actual stars that the object contained.

This third assumption dovetailed quite nicely with the other two major assumptions about the size of the cosmos and the single universe theory. If the universe was small and composed of only the Milky Way, then the nebulae observed, including M31, simply had to be a group of unresolved stars. Get a telescope of sufficient power and it will reveal them to be more stars as part of our galaxy. Yet even later in the nineteenth century, when drawings and photographs of Andromeda revealed its spiral shape, the assumption was that this mass of stars – partially hidden or covered by gas according to Herschel – would and could eventually be resolved to reveal them as a collection of individual stars. That did not happen. In fact, to this day even the best observations of M31 fail to resolve it into individual stars. One cannot even take a telescope and look at the closest star to Earth beyond the Sun, Proxima Centauri, at about 4.2 light-years from us, and resolve it into a globe in the same way we can any of our Solar System's planets into revealing their disks. The reason is simple – these stars are too distant and the same is true for Andromeda and the other nebulae.

The inability to resolve into individual stars because of the vast stellar distances is not something that was considered by astronomers throughout the nineteenth century. Because of assumptions about cosmic distances, because of a belief in a single universe, because bigger and bigger telescopes revealed more stars, observing the Messier or other objects blinded astronomers into believing that the objects that existed were simply more stars yet unresolved.

In some ways they were correct – M31 is a collection of unresolved stars. But it is a collection of unresolved stars that exist distinct from those in the Milky Way. Even as late as 1888, with a photograph of M31 by Isaac Roberts, many still assumed the Andromeda Nebula to be part of the Milky Way Galaxy. This despite a picture clearly revealing its spiral shape that contemporary astronomy associates with galactic structures. Astronomers after the 1920s saw in this picture something earlier astronomers did not see, a distinct galaxy! People often see what they believe to be true. Our senses can trick us into believing Earth stands still and the universe rotates around it, or the world is flat. All of this is a product of the illusion of our senses or beliefs. Copernicus's genius was to shift the paradigm or assumptions about how to think about the universe. Into the nineteenth century there was clearly a paradigm regarding what the universe looked like, and it was ripe and ready to be challenged. Leading that challenge would be new astronomical and scientific tools, and the Andromeda Nebula would be the object of inquiry that would lead that shift in thinking.

Conclusion

Three assumptions guided astronomy into the nineteenth century: First, the universe was of infinite but still definite size far smaller than envisioned today. Second, the cosmos was composed of a single Milky Way Galaxy. Third, deep sky nebulae were simply unresolved stars still part of the Milky Way. These three assumptions had implications for M31. Andromeda was described not as it is today as a galaxy but instead as a nebula. The Andromeda Nebula was to remain the common name for this object at least into the 1920s, when Edwin Hubble's research into galaxies and cosmological distances revealed it to be another spiral galaxy at a great distance (at least by the standards of early twentieth century astronomical knowledge) from Earth.

All nebulae were considered to be part of the Milky Way, including M31. Astronomical research revealed that while many nebulae did have a star at the center, however, for Clerke, nebulae were not necessarily composed of stars but instead were composed of gaseous material, devoid of sun-like bodies. Thus, the nebulae observed were composed of unresolved stars, gas, or both, including the Andromeda Nebula. Finally, because nebulae were part of the Milky Way, their distance was assumed not to be great, at least not exceeding that of the length of our galaxy.

Overall, it would not be unfair to describe the universe of nineteenth-century astronomy as one depicting a finite cosmos of limited size and age, composed singularly of one galaxy, the Milky Way.

4. Andromeda and the Technological Revolution in Astronomy

The Technological Challenge to Nineteenth-Century Astronomy

It is easy to overlook the impact that technology has had on astronomy. Until the invention of the telescope in the seventeenth century, astronomical study was done mostly with the naked eye. The universe was limited to what the visually unaided eye could see. This meant that the night sky consisted of the permanent stars to the sixth magnitude, the planets Mercury, Venus, Mars, Jupiter, and Saturn, the Moon, comets and meteors, and other occasional items such as supernovae. The universe seemed vast to the ancients, but it was clearly small, at least by contemporary standards of the universe, which can see objects to the 25th magnitude, back almost to the beginning of the universe 13.7 billions of years ago.

  Contemporary astronomy stands in contrast. It is not even limited to visible light; scientists can observe in other wavelengths such as ultraviolet and infrared light, seeing the universe in ways that neither the ancients nor individuals in the nineteenth century could imagine. Additionally, astronomy is no longer simply a matter of instant sense impression. Astronomers now have ways to record, preserve, and analyze data. Photography and other technologies have transformed modern astronomy. Finally contemporary astronomy is no longer confined to Earth-based observation. Satellites such as the Hubble Telescope orbit Earth and can record data, and Apollo missions to the Moon brought back rock samples for direct inspection, literally combining astronomy with geology (or lunarology, to be correct, since geology is the study of Earth).

The changes that occurred in astronomy from approximately 1,600 to 1,900 were dramatic. Astronomy went from an observational process rooted in geometry, astrology, and theology to a science of physics, eventually carving out its own unique field of astronomy, or more correctly, astrophysics. Moreover, astronomy was reinvented as new technologies transformed the field. At the center of this transformation was the Andromeda Nebula. As new techniques to study the universe were introduced Andromeda was at center stage.

Although many significant changes in astronomy as a field occurred from 1,600 to 1,800, the 1800s were the critical time. The inventions of photography and spectral analysis made astronomy the field it is today, and it is these technologies, as they were used to study Andromeda, that were critical to this transformation. These technologies, along with rival hypotheses about the structure of the universe and the role of Andromeda in it, set the stage for Edwin Hubble and the major controversies of the twentieth century that rocked our understanding of the universe.

What Is Science?

Today astronomy is considered a science. But that was not always the case. From the ancient times, regardless of whether it was in Europe or elsewhere, astronomy was closely intertwined with astrology, theology, and mythology. Humans turned to the sky and created mythic explanations for why the heavens existed or events occurred. Or they invoked some deities to do the same. Moreover, perhaps one of the earliest reasons the skies were studied were rooted in astrology, with the belief that examining the stars would make it possible to portend and predict the future.

Many still believe in astrology for this purpose and whether for real or fun, still read their daily astrology reports in newspapers or on the Internet. But the astrological interest in foreseeing destinies may have had a practical purpose, too. Astrology and the study of the stars were important for timekeeping. Mastering the passage of time to predict rains or warmer weather for planting was critical to the Egyptians and Native Americans, as well as to

Andromeda and the Technological Revolution in Astronomy 71

other cultures including those in Mesopotamia. The study of the cosmos, then, was wrapped up with many other objectives and forms of inquiry, some of which are not considered sciences, at least by contemporary standards.

What exactly does it mean to be a science? Developing an answer to this question could easily produce a book, since there is a significant amount of dispute over exactly what counts as a science. However, some basic characteristics do stand out.

First, to be a science means the form of inquiry is empirical. An empirical study is one that is based on the gathering of data. Science is not speculative; it is premised on the gathering of evidence. If you want to know something about plants or animals, go look at them. This is what the ancient Greek Aristotle did. He examined scores of animals in order to classify or sort them by certain types of characteristics. Astronomy does that now with stars, based on their spectral analysis, color, size, composition, and temperature. Thus, a starting point for an inquiry about what is a science is that it is about data gathering. If you want to know something, observe and gather data.

Not all forms of observation are scientific. Simply looking at the sky does not make a gaze scientific. More is clearly required. Looking at one star on 1 day is not necessarily a scientific study. If one wants to have scientific knowledge, some type of method of analysis is required. This method is necessary for several reasons. First, if you want to understand the stars or the planets, studying only one of them gives you very limited knowledge. If you looked only at Betelgeuse one would think all stars are red giants; yet this not the case. There are many different types of stars. Observing or examining Betelgeuse might tell you something about this star or red giants, but not about all types of stars. You need to make many observations of perhaps many stars in order to be able to make some broader claims about red giants or stars.

This suggests that scientific learning or an element of the scientific method is that knowledge is based on inductive reasoning. Inductive reasoning is making general statements or conclusions based upon the aggregation or particular observations. For example, if one wishes to be able to make some general statement about stars one needs to observe a lot of them. By studying a lot

FIGURE 4.1 Betelgeuse is the red star in the upper left of the constellation Orion.

of them one might be able to discern general characteristics and therefore make some general statement about them. Studying a lot of stars we might be able to make a general conclusion or claim about their surface temperature or chemical composition. The point is that an aggregation or summation of individual empirical observations allows one to reach broader conclusions about stars. Knowledge is the accumulation of observations or particular facts to reach more general conclusions.

Scientific reasoning thus is inductive. This stands in contrast to deductive reasoning or logic. Deductive reasoning proceeds from general propositions to a particular statement. The classic example of deductive reasoning in logic classes states:

All men are mortal,
Socrates is a man,
Therefore Socrates is mortal.

Although the opening premise might be based on empirical observations, the conclusion here is a simple application of logic. Scientific knowledge is not generally considered deductive since this method is neither empirical nor based on aggregation of individual facts or observations to reach broader conclusions. Astronomy, at least its contemporary form, does not initially proceed by beginning with general claims and then deducing something specific. There may be times there is a deductive aspect to it, though. If once we have concluded a specific star is a red giant we may then infer it has certain characteristics. But initially to conclude something about red giants we need to gather empirical data or evidence in order to make some general claims about them.

Science may be empirical, but sense impressions and observations may not always be accurate. We know our eyes can play tricks on us. For ages humans thought Earth stood still and the Sun and the heavens rotated around us. What we hear, see or observe may be wrong or deceptive. Moreover, what I hear or see might be different from what you perceive. Scientific observation needs to account for the ways our senses trick us. During the Renaissance, individuals, including Francis Bacon and Rene Descartes, argued for rules or methods of inquiry to account for sensory errors. These rules required for example methods of replicating studies so someone else could do the same thing I did. If another followed a certain method of inquiry and reached the same conclusion I did then we should have a result or conclusion in which we are more confident. A strict method meant replication and reliability of results.

Yet another characteristic of the scientific method is the concept of falsification. For a theory to be scientific it must be capable of being tested and ultimately falsified or rejected if the evidence does not support it. The hallmark of scientific knowledge is empirically testing its claims. There has to be some way to test and potentially refute claims if evidence cannot support a proposition. Scientific claims ultimately do not rest on faith, as might the case be with religion. Although many have tried to offer empirical proof or evidence that God exists, ultimately God's existence is an article of faith. This is in contrast to science, which demands proof and the possible testing of its claims. Astronomers seeking to prove the Big Bang claim about the origin of the universe are in search of empirical evidence to support its existence, such as the cosmic microwave background or some type of elementary subatomic particles. Conversely, while searching for this evidence they may also find data that is contrary to claims about the Big Bang thesis. This might include the velocity of the expansion of the universe or the cosmic constant. Theoretically if the evidence is great enough to reject a theory, then it should be abandoned. This is what happened when the heliocentric theory of the universe replaced the geocentric one – the evidence for the latter was lacking and actually proved it to be wrong.

Thus, the core of scientific reasoning is that it is empirical, inductive, it follows rules of a method that allows for replication of results, and it is testable. There are a few additional possible characteristics that might also be part of scientific reasoning. One is that the results are reliable and that they can be preserved and inspected at a later date. By that, while I might claim Martians exist on Mars, if no one else but me has seen them people will be skeptical about my results. Taking a picture or producing other more permanent proof beyond asserting you should take my word for it helps give the claim creditability.

Finally, a full-blown concept of scientific reasoning includes the ability to engage in observation of controlled experiments. By that, if one really wants to know, for example, if a specific drug cures baldness, test it on bald individuals and compare the results to a fake drug or placebo. Controlled experiments are the best way to compare things and determine causes or gather information.

In some scientific areas it is easier than others to do experiments. One can do experiments in chemistry and perhaps in biology or medicine. Yet it is more difficult to do controlled experiments in some areas of astronomy. It is difficult to observe planetary formation and do experiments to see why some form as rocky or gaseous bodies. We cannot create a Big Bang to test a theory, and it might be difficult to explode or contract stars to gather information about supernovae. There are just some limits regarding the use of controlled experiments in astronomy. But even without doing these types of experiments, contemporary astronomy is a science because it is empirical, inductive, method-driven, and its claims or evidence can be preserved, replicated, and tested. But this was not always true.

Transforming Astronomy into a Science

Astronomy has progressed through several stages in its history before arriving at its present form as a science that more appropriately can be called astrophysics. Astronomy was initially mythic; then, with the Greeks and in ancient Mesopotamia, it became rooted in geometry. The geometry of astronomy allowed the Greeks and Mesopotamians to make predictions regarding the size of Earth, the dimensions of the universe, and even for more mundane things such as creating a calendar. Moreover, even through the Middle Ages and into the nineteenth century geometry occupied a major portion of astronomy. Tycho Brahe used it to make stellar distance predictions, and the entire concept of parallax was premised upon geometry and angles to ascertain how far Earth was from celestial objects. Yet, as noted before, these estimates were often seriously and widely inaccurate.

Thus, at least until the early sixteenth century, the science of astronomy was visual and rooted in geometry. The telescope changed all that. It expanded the size of the universe, forced a change to a heliocentric view of the cosmos, and produced a new list of planets and heavenly phenomena. The telescope also helped bring down the Middle Ages by questioning the astronomical basis of the Great Chain of Being. But in many ways, the telescope also did not advance the science of astronomy as much as some might think. Astronomy remained visual, rooted in what the eye could see at that moment. Galileo, Halley, Herschel, and others looked into the telescope and only saw the objects or light that hit at an instant. There was no way to gather more light by looking even longer at an object. What one saw through the telescope remained what one saw, unlike with photography which allowed for increased light gathering beyond a second.

Photographs allowed for more detail, color, contrast, and a fuller depiction of what one saw. Additionally, the only way to record what one saw was by describing objects with words or with hand sketches. These descriptions were good records but subject to errors and mistakes in terms of what an astronomer could write or describe. Unfortunately, this type of astronomy did not allow for others to question what was seen. One had to hope that what one astronomer wrote down or sketched or observed was accurate. For example, Giovanni Schiaparelli (1835–1910), an Italian astronomer, claimed to see canals ("canali" in Italian) on Mars, proof perhaps of an advanced culture on that planet. Yet later on the canals proved to be natural phenomena on the planet. However others, such as Percival Lowell (1855–1916), a businessman and an astronomer who funded the Lowell Observatory in Arizona, was convinced that what Schiaparelli found did prove the existence of intelligent life on Mars, and he spent a career seeking to verify it. The mistakes of Schiaparelli, or at least mistranslations of his descriptions, produced errors in observation and research for others.

Yes, someone else could view the same object, but ultimately, no one could really second guess what one observer saw through a particular 'scope' at a specific day and time. There was thus no really good way to preserve observations. Astronomy, as a science, remained rooted in techniques of recording observations and in prediction that had not really changed much since ancient times.

The move toward really making astronomy a science began with two individuals – Kepler and Newton. Johannes Kepler (1571–1630) was a German mathematician and astronomer. He also was an astrologer, making part of his living by producing horoscopes for fellow students while in school, as well as for friends and his patrons. Kepler was also an assistant to Tycho Brahe. Brahe operated a famous observatory in Denmark, producing a host of scientific instruments that recorded movement in the sky.

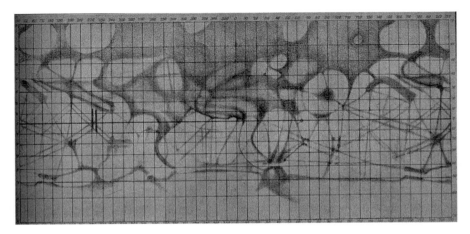

FIGURE 4.2 Schiaparelli's drawing of Mars.

Brahe still believed in the Ptolemaic geocentric concept of the cosmos, and his efforts to use parallax and observations of the sky were directed toward perfecting the model of the universe they relied upon, with celestial objects moving in circles and epicycles around Earth. His observations produced more exact data while also revealing more errors in the current modeling. Brahe sought to perfect the Ptolemaic model but instead his work called it into question.

Brahe brought Kepler on as an assistant in 1600, after the latter had already done some writing and research on the geometry of both the Copernican and Ptolemaic models of the universe. In 1595 Kepler published *Mysterium Cosmographicum* or the *Cosmic Mystery*, which constituted one of the first major defenses of the new Copernican model. In this model the heavenly objects moved in geometrically perfect circles around Earth.

Kepler was fortunate to be assisting Brahe. From him he obtained rich data about the universe, especially of Mars. He took the data and composed it into tables for organizational purposes and analysis. He repeatedly sought to use the data on Mars to construct and perfect a model of its orbit. He also witnessed the famous 1604 supernova; one so bright it could be seen during the day. It was this supernova that he and Brahe used parallax on to determine its distance from Earth. They were generally unsuccessful in their efforts. Yet its appearance was nonetheless significant; it demonstrated that the skies were not permanent, calling into question assumptions about the universe that had prevailed throughout Christendom.

Kepler came of age at a time when the fields of astronomy and astrology were not distinguished, and when physics was considered part of philosophy and math, outside of geometry. It was excluded or not considered a part of the sciences. There was a religious overlay to knowledge, especially in astronomy, with the Christian world order still dependent upon a geocentric universe, and objects moved in fixed circles around Earth. The cosmos was finite and permanent in that there were no new objects that could appear. Finally, it was a universe created by God, with James Ussher (1581–1656), an archbishop for the Church of Ireland, using the Bible to calculate that the universe was created on Sunday, October 23, 4004 B.C. Thus, as Copernicus and then Galileo discovered, any changes or claims that this picture of the cosmos was incorrect met with claims of heresy and threats of excommunication.

The appearance of the supernova thus caused problems for the Catholic Church, but it also was an important event for Kepler. It prompted him to rethink some of his most basic assumptions, including that regarding planetary motion. It was its appearance, as well as his inability to calculate an orbit for Mars, that led him to his Copernican moment in 1605, when he abandoned the belief that the planet orbited the Sun in a circle, opting instead for a model that assumed an ellipse. Assuming that Mars had an elliptical orbit accounted for his data and observations. Based on his observations of Mars, Kepler concluded that all the planets moved in elliptical orbits, with the Sun serving as an axis or focus to the planet. This claim thus formed the first of Kepler's laws. He published this claim and law in his 1609 book *Astronomia Nova* (*New Astronomy*).

In total, Kepler developed three important laws of planetary motion. The first law, as noted above, states that planets orbit the Sun in an elliptical orbit, with the Sun serving as one of the foci. The second law states that planets move faster when closer to the Sun than when farther away, such that if we imagine a line connecting the planet to the Sun, as the planet moves it covers equal areas in equal time periods. This second law requires more explanation.

If planets are in elliptical paths around the Sun then there are some points at which they are closer to it than other points.

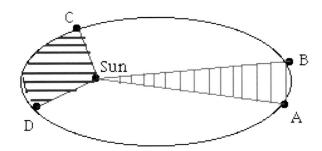

FIGURE 4.3 Kepler's second law of motion.

Now imagine the movement of Mars from points A to B compared to points C to D. Assume a line can be drawn from the Sun to the various points A, B, C, and D. Triangles A, Sun, and B and C, Sun, and D are formed as a result. Both of these triangles are of equal area. More importantly, the time in which the planet moves from A to B is the same as from C to D. The distance from A to B is greater than C to D, but the planet moves more quickly when closer to the Sun and therefore covers more ground. Observationally this is important because it accounts for the apparent speeding up and slowing down and reversal of direction of the planet in the sky.

Because of this change in velocity along the elliptical Kepler could eliminate some of the epicycles from the old Ptolemaic model. Now if both Earth and another planet such as Mars are moving in elliptical orbits at varying distances from the Sun, they will be moving at different velocities at different times. There will be times where Mars seems to move in one direction, slow down, and then move in a retrograde motion. This is not really happening. Instead, Earth may be overtaking and passing Mars in its orbit, at times moving more quickly than the latter. Thus, Kepler's second law can account for this observational phenomena, thereby eliminating the entire framework of cycles and epicycles found in the Ptolemaic geocentric model.

Kepler's third law of planetary motion states that the square of the orbital period of a planet is equal to the cube of its semi-major axis. Mathematically, this third law may be stated as $p^2 = a^3$, where "p" is the orbital period of the planet in years and "a" is the semi-major axis of the planet as measured in astronomical units (AU). What this relationship does is to mathematically make it possible to be able to predict the distance a planet is from the Sun if one knows its orbital period, or vice versa.

Kepler's laws constituted a significant leap in astronomy, making it possible to calculate planetary motions more accurately than anyone previously, while eliminating most of the baggage associated with the Ptolemaic models. The laws were simple and easy to use, and they enhanced predictability, a significant movement toward making astronomy into a modern science.

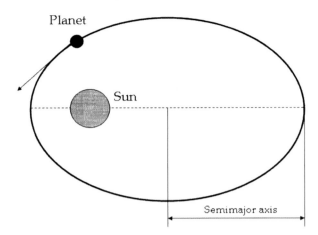

FIGURE 4.4 Kepler's third law of motion.

Isaac Newton

Kepler's three laws of planetary motion are descriptive. They tell us why the planets move the way they do in very accurate and predictive detail. This description is clearly an advance over what previous astronomers, including Ptolemy and Brahe, accomplished. Yet what Kepler did not do was offer an explanation of why the planets move they way they do. This is what Isaac Newton accomplished.

Isaac Newton (1643–1727) was an English mathematician, scientist, astronomer, philosopher and perhaps simply genius extraordinaire. His accomplishments and contributions to astronomy are multiple, but perhaps the most important is his elevation of astronomy to a new level beyond simple observation. His work moved beyond mere description to explanation, giving reasons why specific bodies such as planets move the way they do. Thus, while astronomy up to the time of Newton was based on observation and then description, he moved beyond that to using mathematics to get at the core of why the heavens operate the way they do. The core contribution of Newton was his three laws of motion and, more importantly, his description of gravity.

The first law of motion states that a body at rest or moving in a straight line will continue that way unless acted upon by an outside force. All other things being equal, all bodies and objects including planets will continue to do what they are doing unless enough external force is exercised upon them to bring about a change. A baseball sitting on the ground will stay there until picked up and thrown by a pitcher, for example. Once hit by a batter, it will continue to travel in a specific direction unless some other force alters that. Similarly, an object in space, such as a planet, will stay in motion and move in a constant direction until such time as something else forces it to change.

The second law of motion states that an object's acceleration is proportional to the outside force acting on it. Mathematically this relationship may be stated as:

$$F = ma$$

where F = the new outside force on an object, m = the mass of an object, and a = the acceleration of an object.

Mass is different from the weight of an object. Mass refers to how much material or stuff is in an object. The weight of something reflects the pull of gravity on it. Objects of the same mass subject to different gravitational pulls will be different weights. On the Moon gravity is less, and therefore an individual will weigh less on it than on Earth. Newton's second law of motion can eventually connect mass to weight if the force one is referring to is gravitational pull. However, for the purposes of this discussion, the importance of the second law is simply to establish that once an object is in motion (first law) one can predict its acceleration if we know its mass and the strength of an outside force. Thus, the acceleration of an object in space such as a planet can be calculated if certain things about it are known.

Finally, there is the third law of motion. This law states that for every action there is an equal and opposite reaction. Specifically, when an object strikes another object the second one exerts an equal and opposite force on the first. Thus, the Sun is exerting a force on Earth that keeps it in orbit and in turn Earth is exerting an equal force upon the Sun. However, because of the differences in mass or size of the two objects, Earth moves around the Sun (accelerates) quickly, and that explains in part why the latter circles the former.

Now these three laws of motion tell us basically why objects move, but not fully. This is where the law of gravity fits in. Pop culture depictions of Newton and gravity often show him sitting under an apple tree. He is in deep thought, and an apple falls and hits him on the head. Supposedly this was his inspiration for discovering gravity. This story may simply be a folktale. The other folktale or joke is if Newton discovered gravity, what existed before he found it? Gravity did exist before Newton discovered the math and law behind it.

Prior to Newton gravity did not exist in the sense of an explanation for why objects such as apples fall to the ground. Instead, the argument was made that since Earth was the center of the universe all objects fell to Earth in order to get to the center of the universe. Maybe this argument made sense in a geocentric universe, but it definitely did not in a Copernican one. Since Earth was not at the center of the universe it would be hard to make this argument. Another explanation for falling objects was needed. Enter Newton's theory of gravity.

Simply stated (if that is possible), Newton's theory of gravity asserts that two bodies are attracted to one another with a force proportional to their masses and inversely proportional to the distance between them. Mathematically stated, the law of gravity can be expressed as:

$$F = G(m_1 m_2 / r^2),$$

where

F = the gravitational force between the objects,
M_1 = the mass of the first object
M_2 = the mass of the second object
R = the distance between the two objects, and
G = the universal constant of gravitation

Thus, we can compute the gravitational pull between the Sun and Earth if we know their masses, distances apart, and the universal constant of gravitation (G). Experiments have established a value for G, expressed in Newtons and kilograms as $G = 6.67 \times 10^{-11}$ Newton m²/kg².

Perhaps the simplest way to describe what gravity does with planets is with an analogy. Take a ball, attach a string to it, and

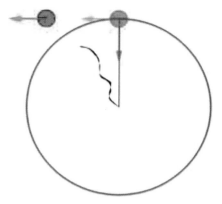

FIGURE 4.5 Ball on a string demonstrating gravitational pull on a planet.

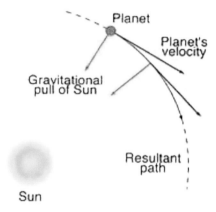

FIGURE 4.6 The gravitational pull of the Sun prevents planets from flying away into space.

start twirling it in a circle. The ball, as shown in the above diagram, has inertia (Newton's First Law) to go straight, but another force (the pull of the string) prevents that from happening, thereby directing the ball to move in a circular path. The greater the pull of the string the more the ball must accelerate to prevent from being pulled out of its orbit and into the string. All this is consistent with Newton's second and third laws.

The same phenomenon occurs with planets revolving around the Sun. The Sun and its gravity act like the string as it seeks to pull the planet (ball) toward the center of the Sun. The planet is already in motion (due to the nature of planetary formation theory), and it would keep moving in a straight line except for the Sun's gravity pulling it inward. Thus, to escape the pull it must accelerate or move quickly enough to maintain an orbit. Hence, while planetary bodies should move in a straight line (Newton's first law) gravity forces the path into a circle, or more correctly, an ellipse.

Newton's description of gravity (this is better than saying discovery) is significant on many scores. First, Newton was able to provide an explanation for why Kepler's descriptions were correct – in particular, Kepler's third law, $p^2 = a^3$. Newton was able to combine his theory of gravity with Kepler's third law to explain the relationship between orbit and length or distance of the semi-major axis. The pull of gravity as determined in part by the mass of the Sun and the planet in question, along with their distance apart, determined the rotational period. Similarly, gravity could also be used to explain Kepler's second law that described the relationship between time and areas of orbital sweep or movement. Newton thus explained mathematically a relationship that Kepler only described.

More powerfully, Newton redefined how to think about the universe with gravity. Prior to Newton God or some other cause (such as an Aristotelian first cause) was invoked to account for motion and movement. A popular theory of philosophy called occasionalism at the time in fact ascribed all causality and movement to God. What Newton did was effectively to remove God from an explanation for movement in the universe. The universe instead was governed by inanimate forces such as gravity that could explain motion. Newton's three laws of motion plus gravity completely destroyed the old ancient world of using myth or religion to explain the cosmos. One could now understand it by way of forces.

The Newtonian model of forces provided a mechanistic view of the universe that remained intact until Albert Einstein in the twentieth century made major modifications in the Newtonian model. But even then, the core ideas about gravity persist and remain valid explanations of the universe. Isaac Newton thus placed astronomy on an even more firm footing in its modern scientific path by the mathematical and mechanistic description of the universe. His 1687 *Principia Mathematica*, where his ideas were published, remains one of the most important books ever published in western science.

Yet Newton's contributions to astronomy and science did not end here. His second major contribution had to do with the principles of light. Until Newton no one really understood light beyond the observation that it was brighter outside with the Sun up than at night when it was not. Moreover, the rainbow, which appeared after rain showers, was a mystery. No one really understood what light was. People were acquainted with what Newton called the "Celebrated Phenomena of Colors," where sunlight passed through a prism to produce a spectrum. Prior to Newton the belief was that the prism added colors to the sunlight. Newton tested this theory by passing light through one prism and then a second. If the theory were correct that prisms add colors then the second prism should have altered the colors from the first even more. It did not. Newton's conclusion was that light was composed of small discrete particles too small to be seen. His theory was important because it suggested light was something that existed and that a prism or the bending of it was what revealed it to be composed of many colors, each of which had its own particles. The importance of this discovery will be discussed later in this chapter.

Around the same time as Newton, Christiaan Huygens proposed light was a wave as an alternative theory. Contemporary physics now asserts that light has both particle and wavelike characteristics.

Newton's interest in optics and light also led to his third major contribution to astronomy, the invention of the Newtonian, or reflector, telescope. The telescope of Galileo was a refractor, and it relied upon lenses for operation. To achieve a more power telescope larger lenses at increasing distance from one another would be needed. This would result in increasingly heavy, large, and bulky telescopes. Newton's reflector used mirrors, and it produced telescopes more compact, powerful, and less costly than refractors. Newton's telescope advanced observational astronomy significantly.

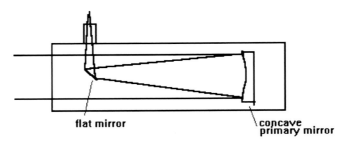

FIGURE 4.7 Refractor telescope.

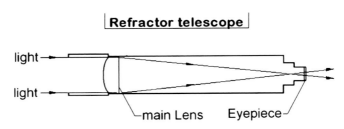

FIGURE 4.8 Reflector telescope.

Post-Newtonian Astronomy

Kepler and Newton transformed astronomy and moved it toward more of a science distinct from astrology and theology. Yet prior to the nineteenth century, they were not the only important innovators or scientists influencing the study of the universe. Many others made important contributions, and only a few can be noted here.

Edmond Halley (1656–1742) was an English mathematician and astronomer most famous for the comet named after him. He argued that the comets appearing in 1456, 1531, 1607, and 1682 were all one comet and predicted this same comet would reappear in 1758. He did not live to see it, but it did, and Halley's Comet has remained the designation since then for it. But perhaps even more significantly, he used the transiting of the Sun by Venus to obtain an accurate account of the distance of Earth to the Sun, and he also was successful in using Ptolemy's tables to compute the proper

FIGURE 4.9 A 1910 image of Halley's Comet.

or real motion of the fixed stars over nearly an 1,800 year period. Halley's major contribution was in combining mathematical and observational astronomy to make accurate predictions of celestial movement.

Johann Bode (1747–1826) was a German astronomer who, along with Johann Titius (1729–1796), noticed that there was a relationship between the planets. Specifically, they saw that between the planets the ratio changed in a fixed pattern. According to the original formulation of the law:

$$a = (n + 4)/10$$

where $n = 0, 3, 6, 12, 24, 48$.

Except for the first two numbers, each subsequent one is double the previous. This ratio almost perfectly captured the increased distance and spacing between the planets. Subsequently, when Uranus was discovered by Herschel in 1781, it fit perfectly into this pattern.

Body	Actual distance (A.U.)	Bode's Law (A.U.)
Mercury	0.39	0.4
Venus	0.72	0.7
Earth	1.00	1.0
Mars	1.52	1.6
		2.8
Jupiter	5.20	5.2
Saturn	9.54	10.0
Uranus	19.19	19.6

Notice here a problem. There is a large gap between Mars and Jupiter. There should be something between Mars and Jupiter but instead there is a gap from 1.6 to 5.2, with Jupiter almost three times as distant from the Sun as is Mars. There should be a planet or something in between, and that led astronomers on a search for another planet.

Giuseppe Piazzi, a Sicilian astronomer became the first to strike gold on January 1, 1801, the first day of the nineteenth century. He claimed to find an object in Taurus. It shifted overnight, and he was convinced he had found the missing planet. He wrote Bode, but by the time the letter reached Bode the object was no longer visible in the night sky. However, a German mathematician, Karl Friedrich Gauss, performed some calculations and argued that it should reappear in Virgo. It was found in that location on December 31, 1801. This calculation by Gauss, as well as Bode's law, signaled the importance of math as a discovery tool in astronomy. This new object was named Ceres by Piazzi, and it was the first asteroid to be discovered. Subsequently, Heinrich Olbers in 1802 found the second asteroid, naming it Pallas.

The discovery of these objects was significant on two scores. First, they confirmed something did exist between Mars and Jupiter, thereby proving the validity of Bodes law. Whether Ceres, Pallas, and the subsequent discovery of other asteroids proved the existence of another planet, though, remains controversial. To this day astronomers are divided. Some contend that a planet never did exist and that the asteroids are the remnants of a planet that failed to form.

According to one theory of planetary science, planets are formed by the accretion of rock and other material as the debris collects around a star such as the Sun. Collision of material along with gravity gradually accretes or combines smaller chunks of material into a planet. Thus, the Asteroid Belt is material never formed into a planet. A contrary hypothesis is that a planet had formed but the gravitational pull of Jupiter broke it apart. Thus the Asteroid Belt is the remnants of a destroyed planet. There is good evidence that even today Jupiter exerts an influence over the Asteroid Belt. There is no need here to resolve whether the belt represents a failed or destroyed planet; instead, it is enough to say that the discovery here of the asteroids was significant. Prior to this, astronomy generally went from observation to theory. Now theory, or at least a mathematical theory, suggested discoveries or phenomena. This was a subtle and important shift in how astronomy operated.

The significance of this shift in astronomy can be found in the search for additional planets. Once Uranus was discovered efforts to map out its orbit were undertaken. However efforts to track it perfectly failed. Some contended that perhaps Newton's laws of motion and gravity were inaccurate. In 1843, John Couch Adams, a British astronomer, hypothesized, consistent with Newton, that perhaps another object was exerting a gravitational pull on Uranus.

FIGURE 4.10 Neptune.

Adams undertook calculations, suggesting that the unknown planet in 1845 had passed Uranus in its orbit. He predicted the planet to be located in Aquarius. He communicated his claim to George Airy, Astronomer Royal of Great Britain, but the latter dismissed the claim. About the same time the French astronomer Joseph Le Verrier came up with calculations similar to Adams. Now Airy was convinced, and a search was launched. Airy convinced James Challis at Cambridge Observatory to search for Planet X while Le Verrier did the same with Johanne Galle at the Berlin Observatory. On September 23, 1846, the Berlin Observatory found the planet, which Le Verrier subsequently named Neptune.

Discovery of Neptune was a mathematical feat. It signaled a new way to do astronomy. It had confirmed Newton's mechanistic view of the universe to be correct, and it showed how astronomy could be theory-driven. In other words, theory suggested what the cosmos should look like, with subsequent observation confirming or testing the theory. This was real science. After Neptune, the discovery of Pluto in 1930 by Clyde Tombaugh at the Lowell Observatory was also predicted by theory, again confirming the power of the new emerging science of astronomy.

The Challenge to the Single Universe Theory

By the beginning of the nineteenth century the prevailing theory still was that the Milky Way was the sole galaxy in the universe and that M31, the Andromeda Nebula, was simply part of the former. Although most nineteenth-century astronomers shared a belief that the Milky Way was unique and that there were no other galaxies, this belief was not unchallenged. Two thinkers raised questions that posed rival views of the universe. Immanuel Kant's 1754 *Universal Natural History and Theory of the Heavens* stood out as a counter to the prevailing thesis that the Milky Way was unique. Similarly, Pierre Laplace (1749–1827), a French astronomer, in his multi-volume (first volume 1799) *Mécanique Céleste* or *Celestial Mechanics* also challenged orthodoxy.

Kant (1724–1804) was one of the greatest philosophers who ever lived. He was born and spent his entire life in Konigsberg, Germany, now the city of Kahlingrad, Russia. His main philosophical work,

The Critique of Pure Reason (1781 and then 1787), is arguably the most important book ever written in philosophy in the west.

In that book Kant seeks to reconcile two trends in philosophy, the belief that all knowledge of the external world is based on sense impressions or simply empiricism, and a contrary theory stating that it is simply through reason that one can gather information about the external world. Roughly described, the latter is rationalism. The problem with empiricism is proof. If all knowledge of the external world is based on sense impressions, what evidence do we have that this is true? How do we know that the ideas in our head conform to the sense impressions we have of external objects? We have no independent experience of viewing ourselves experiencing external objects and being able to verify that what we think is identical to what we saw. Thus, pure empiricism faces a problem. Similarly, pure rationalism raises the problem of getting outside of our head. How do we know the external world if all thought is the product of reason?

Kant, in clear reference and a nod to Copernicus, eventually argues that instead of assuming that all knowledge must conform to external objects, let us assume these objects must conform to knowledge. In effect, we perceive the external world through specific categories or concepts of knowledge. The external world is comprehended via specific categories of human understanding. We cannot ask what the world looks like independent of us viewing it. We cannot answer that question. We must ask what the world looks like to us. The Copernican turn that Kant references here is to say that in the same way Copernicus suggested we think of the Sun and not Earth as the center of the universe, we need not to think of external objects determining human knowledge but instead human categories of knowledge determining how we think about the world.

The implications of Kant's *Critique of Pure Reason* are enormous, and most are beyond the scope of this chapter. Yet one of the most significant was that his arguments made it difficult or almost impossible to make empirical claims or assertions about metaphysics (the origins of the universe) and God. Some claim that Kant's philosophy was the final and definitive blow to the Great Chain of Being and Christian cosmology that had reigned for nearly 2,000 years.

In addition to the *Critique of Pure Reason*, Kant wrote two other critiques that were influential in terms of moral philosophy. His political writings had a major influence in bringing down monarchies, and his thoughts on aesthetics contributed to theories about art. Overall, across the board and in many areas of human knowledge, Kant's influence is significant. Yet in the area of astronomy his 1755 *Universal Natural History and Theory of the Heavens* is profound. The book discusses a wide range of issues, from comets to Saturn's ring, Newtonian mechanics, the cause of the zodiacal light, and the rotation of the planets and the moons. The book also prophetically speculates on planets that might exist beyond Saturn (the book was written 27 years before the discovery of Uranus). In many ways it is a comprehensive review of the state of cosmic knowledge at the midpoint of the eighteenth century.

One of the problems that Kant weighs in on in this book is the origin of the Solar System. Specifically, how was it created, or where did it come from? Kant noticed that all of the planets orbited the Sun in the same plane and moved in the same direction. This could not be a mere coincidence. He asserted that the Sun, Moon, planets, and the comets all had a common origin. They all formed from some similar mass or material, perhaps a large solar nebula. The nebula rotated, and as it cooled, it condensed. The condensation or contraction also produced chunks of material that rotated around what would eventually become the Sun. This material as it cooled and rotated also collided, accreted, and eventually formed the different planets. Thus the Solar System was the product of a nebula forming all the bodies in it.

According to Kant:

> In this system, the development of the planets has this advantage over any other theoretical possibility: the cause of the masses provides simultaneously the cause of the motions and the position of the orbits. Indeed, even the deviations from the greatest precision in this arrangement, as well as the harmonies themselves, are illuminated in an instant. The planets are developed out of particles, which, at the heights where they are suspended, have precise movements in circular orbits. Thus, the masses formed by their combination will continue exactly the same movements at precisely the same level and in exactly the same direction. This is sufficient to understand why the movement of the planets is approximately circular and why their orbits are on a single plane. Moreover, they

would be exactly circular if the distance from which they gather the elements for their development were very small and thus if the difference in their movements were very insignificant. But because the development of a thick planetary cluster involves a wider surrounding area, throughout which the fine basic stuff is scattered so much in celestial space, the difference in the distances of these elements from the Sun and thus also the difference in their velocities are no longer insignificant. As a result, given this difference in the movements, it would be necessary, in order to maintain on the planet an equilibrium between the central forces and the circular velocity, for the particles which collide with the planet from different distances and with different motions to offset each other's aberrations exactly. Although this, in fact, occurs fairly accurately, nonetheless, this compensation falls somewhat short of perfection and brings the deviations from circular movement and eccentricity with it. It is just as easy to shed light on the fact that although the orbits of all planets should properly be in one plane, nevertheless in this part we also come across a small deviation, because, as already discussed, the elementary particles which find themselves as close as possible to the general plane of their movements nevertheless take up some space on either side of it. It would be only too fortunate a coincidence if all the planets were to begin to develop exactly in the middle between these two sides on the plane connecting them, something which would already cause some inclination of their orbits towards each other, although the impulse of the particles from both sides would restrict this deviation as much as possible, allowing it only within narrow limits.

However, if our Solar System formed out of a nebula, is it not possible that around other stars similar phenomena occurred? Kant thought yes, but he expanded his point even further to speculate about not just other solar systems but also other galaxies.

> If, then, the fixed stars constitute a system whose extent is determined by the sphere of the attraction of that body which is situated in the centre, shall there not have arisen more Solar Systems and, so to speak, more Milky Ways, which have been produced in the boundless field of space? We have beheld with astonishment figures in the heavens which are nothing else than such systems of fixed stars confined to a common plane – Milky Ways, if I may so express myself, which in their different positions to the eye, present elliptical forms with a glimmer that is weakened in proportion to their infinite distance. They are systems, so to speak, an infinite number of times infinitely greater diameter than the diameter of the Solar System. But undoubtedly they have arisen

Figure 4.11 Immanuel Kant.

in the same way, have been arranged and regulated by the same causes, and preserve themselves in their constitution by a mechanism similar to that which rules our own system.

Kant's theory, which Edwin Hubble later referred to as the theory of "island-universes," stood in contrast to the single-universe hypothesis that saw the Milky Way as the only galaxy in the universe. Kant proposed that the universe is composed of many other solar systems and potentially many other planets within a cosmos of many galaxies beyond the Milky Way. Kant even refers to the Andromeda Nebula in his discussion but does not offer any speculation on whether it is a galaxy.

At approximately the same time that Kant was proposing his theories about the origins of the Solar System and island-universes, Pierre-Simon Laplace (1749–1827), a French mathematician and astronomer, similarly asserted that the Solar System was formed by way of contraction and accretion. He asserted also like Kant that many of the nebulae observed might be distinct galaxies from the Milky Way, and he speculated that some stars might get so heavy that they would collapse on themselves. The latter comment seemed to anticipate theories about black holes that arose in the twentieth century.

Andromeda and the Technological Revolution in Astronomy 95

FIGURE 4.12 Pierre-Simon Laplace.

The arguments that Kant and Laplace made about the origin of the Solar System have come to be accepted as accurate. Referred to the Kant-Laplace theory, it states that nebula cooling, rotating, and eventually collapsing and accreting form stars, planets, and the other objects around them. Moreover, extrapolating from Earth's Solar System, Kant and Laplace turned out to be correct in their assertions that other galaxies existed and that the Milky Way was not the sole galaxy. Finally, as the discovery of exoplanets has demonstrated, Kant was also correct in guessing that other worlds may exist.

The Kantian theory about other galaxies posed a challenge to astronomical orthodoxy well into the nineteenth and twentieth century. However, in order to provide evidence for this argument, two other dogmas or accepted truths would also have to be challenged. The first would be to argue that the distances in the universe were much greater than previous thought, such that Andromeda would have to be thought of as being further away from Earth than current measurement suggested. The second would be to indicate that not all nebulae were unresolved stars and that there were differences among the different objects gene-rally classified by Herschel and others as nebulae. Two inventions would help dramatically to resolve these issues – photography and spectroscopy.

Astronomy at the Beginning of the Nineteenth Century

The telescope opened up new worlds – such as with Galileo first seeing the moons of Jupiter – affecting a Copernican revolution in how we understand the heavens and our role in it. Yet even as the telescope transformed our knowledge of the universe, one aspect of astronomical science remained constant: It was a science of instant impression. Astronomers could look through larger and larger telescopes, yet what they saw in them was dependent upon observations that lasted literally only as long as a blink of an eye. Observers could do drawings, but they were at best composites of what astronomers thought they saw in the telescope. These drawings failed to provide astronomers with the opportunity to double check what others had seen. So long as replication and examination of what others had actually seen (a hallmark of the scientific method) was impossible, astronomy could not evolve into a science. Missing was a permanent record of what was actually seen.

Photography transformed human history in numerous ways. Invented in the early to middle part of the nineteenth century by Frenchman Louis Daguerre (1787–1851) as well as others, photography allowed for the creation of permanent and exact reproductions of objects that we see. Instead of portraits of people photos became possible. Photography made it possible for one person to take a picture and share it with others. Instead of a hand drawing or verbal or oral description – all of which faced problems in terms of accuracy of representation – a photograph let others see what you saw. Others could question or verify observations. The importance of photography was not lost on astronomers.

In some ways, photography and astronomy are very much related, and joining them together to create astrophotography made a lot of sense. Telescopes, especially reflectors and their mirrors, were based on light-gathering power. The bigger the mirror the more light it could gather and therefore the more distant objects could be seen. Roughly speaking, more light gathering meant greater magnification or more powerful telescopes that enhanced observation by enlarging objects or bringing new ones into the view that could not previously be seen. Photography worked on a premise that as light struck a specially coated surface, such as one with a silver compound, as was true with early cameras, this light would capture the image of the object being photographed. This image could be captured on a photographic plate and once developed, it would be possible to have a permanent image of that object.

In 1839 Daguerre unsuccessfully sought to photograph the Moon, producing a blurred image. But a year later John William Draper (1811–1882) produced the first photo of the Moon, shooting it through a small reflector telescope. After Draper, other astronomers experimented with combining telescopes and cameras to produce pictures. The boon to astronomy was terrific. It allowed for the production of permanent records of objects seen and, importantly, with records preserved, one could make comparisons of the sky over several nights to detect objects that were moving in relation to other objects. This is what Clyde Tombaugh did in the 1930s to discover Pluto, and this technique has also been employed to locate other objects such as asteroids, comets, and Kuiper Belt objects. Now with photographs of the sky dating back to the nineteenth century, it is possible for astronomers to look back 150–200 years to see how the sky has changed.

Astrophotography was also aided by the invention of equatorial devices that allowed telescopes to follow objects in the sky as they moved due to Earth's rotation. The first crude equatorial telescopes date back to Galileo's time, but it was not until the eighteenth century that that became more prevalent. Equatorial-mounted telescopes made it easier to follow the path of stars in the sky by coordinating the observer's location (latitude) along with the specific coordinates of a star or object in the sky. Once an object is located one merely needs to keep moving the telescope in order to stay on focus. Today there are computers and motors that allow for that tracking, but in the nineteenth century equatorial telescopes necessitated hand tracking or movement. The point here is that as telescope design and tracking improved, it allowed for longer and longer astrophotographs in terms of time. The longer the lens of a camera attached to a telescope was open, the more light could be gathered. This meant fainter objects could be seen, or those already detected could be resolved or viewed more clearly as more detail was obtained. This detail included colors (once color photography was possible) and nebulae details, perhaps showing or resolving individual stars.

98 The Andromeda Galaxy and the Rise of Modern Astronomy

FIGURE 4.13 Isaac Roberts' 1888 photograph of M31.

Prior to astrophotography, efforts to capture or depict the Andromeda Nebula were still done with hand drawings aided by the telescope. In 1871 Lord Rosse produced one of the first hand drawings of M31 that seemed to show the central region of the object. He produced this figure through observations with his 6-foot reflector telescope. The drawing reveals a central core and also what appears to be a spiral design, but it is not clear that the object seen is composed of individual stars.

Not a surprise, one of the first significant astrophotographs taken was that of the Andromeda Nebula. Isaac Roberts in 1888 produced a photograph of M31 that revealed its spiral shape, but the individual stars could not be resolved, reinforcing the notion that the Andromeda really was a nebula. Not until the 1940s would the stars of M31 be resolved, providing valuable evidence of its galactic nature. The picture of Andromeda was significant. It definitely looks similar to many of the pictures seen today, revealing the spiral and bright central core. But still this did not provide conclusive evidence about its distance, whether it was distinct from the Milky Way, and what its nature or composition was. Facilitating answers to these questions would require another technology – spectroscopy.

Astronomy, Light, and Spectroscopy

Even with a telescope and a camera astronomy is still a distant observational science. Astronomers can look at distant objects but not really know much more about them beyond speculation and educated guesses. Maybe the Moon is made of rock or green cheese – by simply looking at it we don't know.

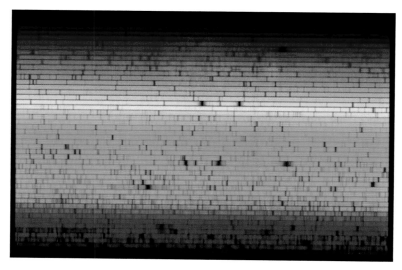

FIGURE 4.14 Spectrograph of the Sun.

In 1814 Joseph von Fraunhofer (1787–1826), a German optician, repeated Isaac Newton's famous sunlight and prism experiment. Yet this time Fraunhofer magnified the resulting rainbow, discovering more than 600 lines across the rainbow. Today these lines are known as spectral lines, with estimates being that sunlight has more than 30,000 lines. Exactly what these lines were was unknown until 1857.

In that year the inventor of the Bunsen burner, Robert Bunsen, a German chemist (1811–1899), decided to experiment with burning different substances in a gas burner. Chemists had known for years that when materials were burned they changed the color of the flame. Bunsen invented his burner to facilitate the study of why these materials changed color when burned. His assistant, Gustav Kirchoff, then suggested that the light be examined after passing it through a prism. They too found that the spectrum produced spectral lines. But these were bright lines against a dark background. In fact, they noticed patterns where different elements or materials yielded different line patterns. They connected the lines seen by Fraunhofer with theirs, concluding that each element produced its own unique lines, almost a DNA for each one. From his experiments he devised three rules now known as Kirchhoff's laws:

- A hot body or a hot dense gas produces a rainbow spectrum without any spectral lines.
- A hot gas produces bright spectral lines against a dark background.
- A cool gas produces dark spectral lines against a rainbow spectrum.

The bright lines have come to be known as emission lines and the dark ones absorption lines. Moreover, when the same element is burned, the emission and absorption lines appear in the same location, again reinforcing the idea that the two types of lines are the same.

Why is all this significant? Let's return for a moment to Newton.

Newton thought light to be a series of small particles. Conversely, Huygens asserted it to be more like a wave. Experiments undertaken by English scientist Thomas Young in 1801 revealed light to have wavelike traits.

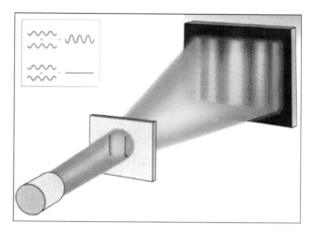

FIGURE 4.15 Thomas Young's light experiment.

Young passed light through parallel slits, forming bright and dark patterns. If Newton were correct the images formed would be of the slits. Instead this bright and dark pattern meant that light was bending like a wave. Why this wave pattern was significant was resolved in the 1860s when James Maxwell (1831–1879), a Scottish physicist, began experiments on electricity and magnetism. He concluded that the two were linked together into one force called electromagnetism. Moreover, during his experiment and in his mathematical calculations he determined that the two forces traveled at the same speed as light -3.0×10^5 km/s, or approximately 186,000 mile per second. The conclusion Maxwell reached? Light is a form of energy, a form of electromagnetic radiation.

This conclusion is significant for astronomy. If light is a form of energy and a spectrum of a burning object has unique spectral lines, then it is possible to determine the chemical composition of light. Thus, one should be able to shine any light through a prism, observe its spectral lines, and therefore determine what its chemical composition is. One could divert light from the Sun, Moon, planets, and even distant stars and stellar objects, all with the aim of being able to ascertain their chemical composition. The astronomical implications here were enormous.

Think about what happens if you combine spectroscopy – which made possible analysis of the chemical content of distant phenomena – with a telescope and photography. Aim a telescope at a distant object, do a spectral analysis, and then determine its chemical composition. Now combine that with the camera and a permanent record is produced.

The unity of the three technologies would forever change astronomy and usher in the field of modern astrophysics. Merging spectroscopy and photography meant that permanent records of what was seen could be captured and preserved and that exposures were far longer than the bat of an eye. Instead, astronomers could now see even more details than before, while allowing others to review what others had seen. Overall, introducing new technologies to astronomy opened up even more avenues of knowledge and techniques for studying the sky. Astronomers did not simply have to speculate about what distant objects were; they could now do spectral analysis and find out.

At the center of this merger of astronomy and these new technologies was the Andromeda Nebula. Among the first objects studied or examined with a spectroscope was M31. William Huggins (1824–1910), a British pioneer in amateur astronomy and spectroscopy, compared different types of nebulae, including Andromeda. He noted as early as 1864 a difference between the Orion Nebula, which displayed an emission spectra characteristic of other gaseous nebulae, and Andromeda, which produced a spectrum more typical of stars. This conclusion reinforced claims that Andromeda was composed of unresolved stars, yet it did not lead to the assertion that it was a separate galaxy.

A second and more significant spectral analysis of Andromeda was published in 1899 in the *Astrophysics Journal*. Julius Scheiner's brief piece reported an analysis of a seven-and one-half hour photograph he took of M31's spectrum. Although he did not publish the photograph, Scheiner reported that the continuous spectrum of Andromeda was similar to that of the solar spectrum (our Sun). His analysis seemed to confirm Andromeda as a collection of stars, yet in contrasting spiral nebulae (such as Andromeda) from ring nebulae (the former with stars, the latter not), he again did not reach the conclusion that M31 was a distinct galaxy. Instead, while noting some differences between stars in the Milky Way from those in Andromeda, he did not venture a hypothesis on the distinct galactic structure of the latter.

Summary

The nineteenth century witnessed a dramatic change in astronomy. Although the telescope and math had been important to astronomical research prior to that century, the perfection of both combined with the new technologies of photography and spectroscopy moved astronomy in a new direction. It fully emerged from being linked with astrology and theology, achieving much more of a scientific rigor than it had prior to the nineteenth century. Astronomy moved from an observational field based on instant impression to where theory drove discovery and where distant objects could be examined scientifically in ways not possible before that time. Figuratively and literally astronomers no longer were confined to looking at distant objects but they could also reach out and render conclusions about the chemical properties of the cosmos. Permanent records could be preserved via photography, and this made it possible for astronomers to test and verify observations. In short, astronomy emerged as a science.

Leading the way in the move toward astronomical science was the Andromeda Nebula. One of the first objects scientists turned to when telescopes were first invented was Andromeda. The same was true when astrophotography and spectroscopy were developed. Astronomers quickly turned to M31, seeking to find out what it was or what its composition was. The study of Andromeda proceeded along with the rise of new astronomical technologies and the field of astronomy as it became more scientific and more based on physics and math. The difference between astronomy at the beginning and the end of the nineteenth centuries was that at the beginning the field looked more like astrology, by the end it was more like astrophysics – the name or label ascribed to the field today, reflecting its merger or close relationship with physics. It represented a field premised on the laws of Kepler and Newton, a mechanical universe of forces.

In studying Andromeda, astronomers looked for basic answers about the universe. They sought to determine what it was made of and why it was structured the way it was. The prevailing wisdom of nineteenth-century astronomy remained one of a single galaxy universe of fixed dimensions, but theories by Kant and Laplace, and spectral and photographic analysis began raising questions about this view of the cosmos. By 1900, although orthodoxy prevailed, many forces were in play that set the ground to question astronomical dogmas. Describing those forces is the subject of the next chapter.

5. Andromeda and Astronomy at the Beginning of the Twentieth Century

Astronomy traveled the distance of a galaxy in the nineteenth century. It entered the century as a study premised upon the impressionistic, snapshot observations by the human senses, while exiting it far more a science. Many forces, building from the time of Copernicus, Kepler, and Newton contributed to that transformation, including the adoption of mathematics as a predictive tool and the increased reliance upon empirical evidence and data to support cosmological claims. Moreover, the legacy from Copernicus – not necessarily to trust what your senses tell you but instead seek what makes explanatory sense – also drove the field of astronomy into the nineteenth century.

Photography and spectroscopy, however, were the major technological forces that remade astronomy during the nineteenth century. They allowed for astronomy to become astrophysics – the application of the laws of physics to the study of the heavens. These two technologies also made it possible to move astronomy into the realm of preserving observations, thereby rendering observations more permanent and verifiable, and also it had the capacity metaphorically to move humans away from Earth and give them close observations of other bodies. If one could actually use spectroscopy to determine the chemical composition of distant objects, it was like moving the field of geology to the Sun, the Moon, planets, and distant stars. More powerful telescopes brought more objects into view and made the universe seem closer, but spectroscopy moved humans closer to distant objects. We could not really touch and feel them, but we certainly could inspect them as if we were up close.

Despite these changes in technology that rendered astronomy more of a science than it was when it entered the century, by about 1900 many assumptions still held sway. The assumption still was one of a finite, single-galaxy theory composed only of the Milky Way. Distant objects such as the Andromeda Nebula were part of the Milky Way. They were not too far away, and overall distances in the universe were quite small compared to what astronomers think now. Although some thinkers such as Kant and Laplace offered rival interpretations of the universe as composed of many galaxies, orthodoxy was that the universe was a singular universe composed of the Milky Way. Earth may no longer have been viewed as the center of the universe, but the belief still was that there was a center and that perhaps the Sun was its focal point. Earth and humans may not be at the apex of the universe, but an apex did exist, and we were at least close to it.

However, the technological changes of the nineteenth century had snowball effects. New tools of science made it possible to see the universe in a different way. They helped to affect a paradigm change about the universe that would be every bit as significant as what Copernicus and Galileo and the telescope accomplished. Among the forces driving change would be how astrophotography and spectroscopy affected and changed astronomy, ushering in new discoveries and the classification of stars and universe. Second, these technologies, especially in the hands of Henrietta Levitt, made it possible to recalculate astronomical distances. Third, these changes prompted scientists and physicists such as Albert Einstein to rethink the universe. Finally, astronomical observations of Andromeda drove scientists to ask important questions about the cosmos, prompting a revolution in claims about the universe that would culminate in a rejection of the island-universe theory by the 1920s.

The Nature of Light

Three initial claims about light were important as the nineteenth century came to a close. First, the visible light we see is a form of energy. This was the conclusion of James Maxwell. Second, light demonstrates both particle and wave-like traits. This is what Isaac

Newton and Thomas Young showed. Third, the spectra of luminous objects produced specific lines that revealed their chemical composition much like a DNA. This is what the experiments of Bunsen, Kirchoff, and Fraunhofer demonstrated. All three assertions were critically and empirically interconnected in how they described light. Together, these three claims about light along with the tool of spectroscopy made it possible to expand astronomical knowledge in a variety of ways. Let's begin first with the concept of light as energy.

Visible light is energy. Most of us recognize this intuitively. The warmth of the light produced by the Sun on the skin in the summer often feels great, especially after a long, cold winter. The light of the Sun can tan skin. Project the light of the Sun through a magnifying glass and one can burn paper. Taken further, when really hot, one can cook an egg on the hood of a car. Moreover, solar panels generate electricity from the light of the Sun. All of these instances demonstrate that light, heat, and energy are related in some way. But there is a deeper significance to concluding light is energy – it is connected to other forms of energy.

Visible light is what we normally mean we discuss light. Light is what is received from the Sun – a light bulb, a fire, or some other source of illumination. It is the type of light we can see with the naked eye. But light and energy are more than visible light. James Maxwell's experiments examined magnetism and electricity. His work indicated that the two were related and demonstrated that they were a combined force that he called electromagnetism. But Maxwell's experiments also yielded a few additional facts. One was that electricity and magnets generated fields of energy that oscillated out from the source. That is, there were little force fields emanating from each; the closer to the source, the stronger the fields. Second, when seeking to measure the fields Maxwell discovered that the fields pulsated or moved in waves. Timing the waves he found that they roughly moved at what was then calculated to be the speed of light. Thus, Maxwell concluded that electromagnetism was both a form of radiation and related to light. Visible light was thus a form of electromagnetic radiation.

When most people think of radiation, they think of something radioactive or glowing, such as uranium. Uranium is radioactive, and in popular images, it will set off a Geiger counter. But radiation is not simply something that is radioactive. To say light is radiation means only that it radiates – it emits energy in the form of waves. Thomas Young's experiments demonstrating light to have wave-like characteristics meant that it emitted energy according to a certain type of wavelength. Think of light as a piece of rope. Grab one end of the rope and shake it up and down slowly. Notice how it creates a wave. Now shake it up and down more rapidly and it creates a wave, but this wave looks differently than the waves from first shaking. The first one might produce a higher wave, while the latter lower but more frequent waves. There seems to be a relationship between the speed or velocity of the waves and the distance between the waves. Heinrich Hertz (1857–1894), a nineteenth-century German physicist, noticed that also and devised a mathematical formula that expressed the relationship between the two and with the speed of light.

$$V = c/\lambda$$

where v = the frequency of a wave, c = the speed of light, and λ = the wavelength of the wave.

Frequency in this formula is measured in cycle per second and is called Hertz (Hz). The speed of light is 3×10^8 m/s, and the Greek letter λ (lambda) is the wavelength expressed in meters. What is critical about this relationship is that visible light is a form of energy – electromagnetic energy – radiating or oscillating at a specific frequency. There are other forms of energy that also exist, but they are at different frequencies. For visible light, violet radiates at 400 nm/s while red does the same at 700 nm/s. And visible light is only a small part of the broader electromagnetic spectrum.

Visible Light and the Electromagnetic Spectrum

There are many forms of electromagnetic energy radiating at different HZ or wavelengths. These forms of energy run the range from very short gamma rays through X-rays to visible light to microwave and then to radio waves. Light thus emits energy in different wavelengths. What we see with our eyes is only one form of energy.

Andromeda and Astronomy at the Beginning   109

Our eyes operate like visible light detectors. The same is true for telescopes – both are able to capture and detect visible light.

If visible light is only one form of energy that emits from some object, perhaps celestial bodies also project or emit other forms of energy, and if the appropriate detector could be constructed, these other forms of energy could also be detected. The relationship among energy, wavelengths, and eventually the color and temperature of objects expresses another critical set of relationships involving light that were discovered during the nineteenth century and which became important to astronomy.

Let's begin with a simple relationship between cold and hot objects. Part of what makes an object hot is its temperature. But temperature is also related to wavelength and frequency. Specifically, in everyday language, what makes an object feel hot is that it emits heat. Cold objects feel cold because they do not emit heat. Now think of a light bulb. When turned off it is cold and does not glow. Give it electricity and it gets hotter and it glows. Theoretically, heat it up even more and it would glow more or produce more light and heat. There is definitely a relationship here. The same is true with a fire. Really hot fires on kitchen stoves burn with a blue color, whereas fires burning wood, for example, are more yellow or orange. This is because they are burning at different temperatures and emitting different levels of energy. Now apply all this to astronomy and the concept of blackbodies.

What is a blackbody? On one level it is a theoretical concept. It is a body that does not emit any radiation or light, except due to its temperature. Instead, a blackbody object absorbs radiation. Ordinary objects around us such as people, tables, and chairs are forms of blackbodies because they reflect light. More importantly, stars such as the Sun are nearly perfect blackbodies because they too absorb almost all of the radiation around them. The energy or light they emit is a product of their temperature, with the energy emitted referred to as blackbody radiation.

Blackbody radiation can tell us a significant amount about stars. Specifically, it can provide information about temperature. How? Recall that different colors such as red or violet vibrate at different wavelengths. Moreover, the intensity of sunlight varies with its wavelength, with more intense stars vibrating or emitting at a larger frequency. What determines the different wavelengths or intensity is the temperature of the body. The greater the temperature, the greater the intensity, and vice versa. As stars such as the Sun burn more intensely, their temperature increases.

Simply stated, the hotter a star is, the more it glows. This is no different than a fire burning different objects. A fire burning wood is cooler than one burning gas. Moreover, as the temperature of a fire changes, so does its color. Cooler fires are more reddish or orange, hotter ones bluer. The same is true with stars. The hotter they are, or the more intense they burn, the more they will shift from burning red to blue. Thus, by examining the light from a star one can reach some conclusions about its temperature or intensity. Combine that with a spectral analysis and one can then reach conclusions about temperature and chemical composition. The light from a distant star thus can be studied, revealing a significant amount about it.

But the blackbody properties of objects such as stars can also be useful to astronomers in other ways. Nineteenth-century astronomers constructed two formulae to describe relationships that connected temperature, light, and energy. Wilhelm Wein (1864–1928) was a German physicist who constructed a formula

FIGURE 5.1 Wilhelm Wein.

in 1893 expressing the relationship between wavelengths and temperature. The formula was quite simple:

$$\lambda_{max} = 0.0029 / T$$

where λ_{max} = the wavelength of emission expressed in meters, T = the temperature expressed in Kelvin, and 0.0029 = a constant devised by Wein.

What Wien's law demonstrates is an inverse relationship between temperature and wavelength. Double the temperature of a blackbody such as a star and its wavelength is cut or reduced by half. Wein's law is useful to determining the temperature of a star.

But Josef Stefan (1835–1893) and Ludwig Boltzmann (1844–1906), both Austrian physicists, developed a formula to calculate the total amount of energy that a blackbody such as a star emitted or radiated. First off, energy is generally expressed in Joules (J). It is the amount of energy found in moving a 2 kg (roughly 4.5 lb) object 1 m/s. This unit of energy is named after the nineteenth-century British physicist James Joule. Wein and Boltzmann in the later 1870s and early 1880s contended that the total amount of energy emitted by a blackbody was proportional to its temperature and surface area. The hotter and bigger an object, the more energy it radiated. The total amount of energy emitted is referred to as an energy flux. What has come to called the Stefan-Boltzmann blackbody law was born.

$$F = \sigma T^4$$

where F = total energy flux emitted, expressed in joules per square meter per second of surface area; T = the temperature of the blackbody object (in Kelvin); and σ = a constant of 5.67×10^{-8} W m^{-2} K^{-4}.

In the constant σ, W refers to watts, another unit of energy. Commonly, watts are used to describe the energy used by a light bulb. One watt is equal to 1 J/s. This means a 50 W bulb uses 50 J/s.

The Stefan-Boltzmann law describes what happens when either the temperature or surface area of a blackbody (star) increases or decreases. If the temperature doubles or increases by a factor of one (it doubles), then the energy flux increases by a factor of 2^4, or

FIGURE 5.2 Josef Stefan.

FIGURE 5.3 Ludwig Boltzmann.

16 times. This law can be used on any star, not just the Sun, and it then becomes useful for comparing bodies to determine their size and energy emission.

The information about stars as blackbodies and understanding their energy emission and wavelengths has also proved useful in other ways, too. We can begin with color and temperature. Why are

Andromeda and Astronomy at the Beginning 113

Class	Temperature (kelvins)	Conventional color
O	≥ 33,000 K	blue
B	10,000–30,000 K	blue to blue white
A	7,500–10,000 K	white
F	6,000–7,500 K	yellowish white
G	5,200–6,000 K	yellow
K	3,700–5,200 K	orange
M	≤ 3,700 K	red

FIGURE 5.4 Stellar color-temperature relationship.

stars different colors? Looking at the night sky, one sees red stars such as Betelgeuse in Orion, a blue-white star Sirius in Canis Major, a gold and blue double star Alberio in Cygnus, and a yellow star, or our Sun, during the day. Colors of stars reflect the surface temperature of a star. Each temperature indicates a specific wavelength; more exactly, the color of a star represents the peak wavelength that a star emits energy. If a star is hot it peaks at a shorter wavelength, and its color will be closer to the blue range. If it is cooler the wavelength is longer, and it will tend to move toward the red part of the spectrum. Color thus reveals surface temperatures for stars. This is useful to astronomers seeking to understand more about them. Color and temperature permits their classification or grouping by star types.

Based on research done at Harvard in the later 1890s, sponsored by Henry Draper, who, in 1872, became the first to take a photograph of a star's absorption lines, astronomers have classified stars with letters, using a range that moves from the hottest (blue) to the coolest (red). Traditionally the classification has had seven types of stars, employing the letters OBAFGKM. Mnemonic devices to understand this classification have been created, among them: "Oh Boy A F Grade Kills Me" and "Oh Boy A Fine Girl (or Guy), Kiss Me!" More recently, even cooler stars, including brown dwarfs, have been found, necessitating now up to ten different stars, yielding OBAFGKMLTY. What mnemonic device can help with learning this sequence? How about "Onions, Bacon, And Fried Green Kale Make Liver Taste Yummy?"

In addition to color revealing temperature for classification purposes, a spectral analysis reveals the chemical composition of the star. This chemical composition is significant because it helps in part to explain why stars burn at different temperatures. Simply put, they are using different fuels. Spectral lines correspond to the presence of specific chemicals present in stars.

All stars burn and generate energy based on nuclear fusion involving the elements hydrogen and helium. But some stars have more complex elements in them, burning them as fuel. Thus, the hotter they burn due to the fuel they are burning, the brighter they shine.

Now, the varying brightness can also address a couple of other issues regarding stars – the concept of luminosity, size, and then beyond that, the distance of them from Earth. Some stars appear brighter in the sky than others. Measurement of this brightness refers to the relative magnitude of a star in the sky. Astronomers use magnitude to classify the relative brightness of objects, with smaller numbers indicating something is brighter than another entity having a larger number. The Sun is listed at −26.7 magnitude, a full Moon at −12.6, and Sirius, the brightest star, at −1.4. The human eye can generally see stellar objects (or any objects in the sky) to about 6th magnitude, with the most powerful Earth-based telescopes able to detect objects to 21 magnitude. The Hubble Telescope can see to around 30 magnitude.

The magnitude scale is not exactly linear in the sense that an object of magnitude one is twice as bright as a two, etc. The scale is closer to an object of one magnitude being approximately 2.512 times as bright as one whole magnitude below it. A first magnitude star is thus 2.512 times as bright as a second magnitude, a second is similarly in ratio to a third magnitude star. Comparing a first to a fifth magnitude means the former is 2.512^5 or 100 times brighter. Magnitude thus is another way to classify stars.

Relative magnitudes show how stars relate to each other in brightness and not absolute differences in brightness. Absolute differences in brightness could perhaps be calculated if stars were similar in terms of their size and distance from Earth, among other factors. But there are many reasons why some stars are brighter than others. Temperature is one factor, the surface area or overall sizes of stars another, and distances from Earth, too, are important. Yet the apparent brightness of a star can provide critical information about both its distance and what astronomers call it luminosity. Luminosity refers to how much light energy stars emit per second. It is one way to hold constant how bright a star really is. One equation to classify brightness compares or contrasts it to luminosity. An equation expresses that relationship:

$$L = 4\pi R^2 \sigma T^4$$

with L = luminosity, as expressed in watts, R = radius of a star, σ = Stefan-Boltzmann constant, and T = surface temperature of a star in Kelvin.

This equation calculates brightness by considering the surface area or how big a star is along with how high its temperature is. Together the two characteristics tell us something about how much energy the star is producing and therefore how luminous or bright it is. All things being equal, the bigger and hotter a star is, the more energy it produces and therefore the brighter it will be.

Stars thus can be classified in a variety of ways based on temperature, color, spectral type, and size. Is there any way to bring all of these factors together to understand stars and organize them for the purposes of study and analysis? Two late nineteenth and early twentieth century astronomers detected a pattern among the stars that included all of these factors. Danish astronomer Ejnar Hertzsprung (1873–1967) first noted by way of a graph in 1911 that there was a connection between luminosity and surface temperature. Then American Henry Norris Russell (1877–1957) graphed spectral types to surface temperatures, noting a similar pattern. What they produced has since come to be known as the Hertzsprung-Russell, or H-R diagram.

The H-R chart brings together a variety of insights and discoveries about stars by way of classifying them. It connects together temperature, spectral type, and color. But what it also does is bring in stellar size when classifying. Moreover, stars are classified in terms of whether they are a main sequence or part of the OBAF-GKM grouping, and whether they are giants or super giant stars or white dwarfs. Main sequence stars are like our Sun, a G-type

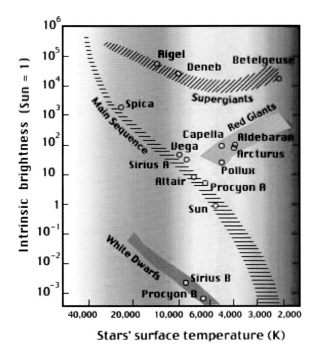

FIGURE 5.5 Hertzsprung-Russell diagram.

star. These are typical stars and are so grouped based on their size, color, and spectral composition. The three factors come together to produce a specific temperature and luminosity for each type. About 90% of all stars are main sequence ones.

Another 1% of stars are either giants, typically red stars such as Arcturus, or supergiants such as Betelgeuse. These stars are curiosities because they are not that hot – about 3,000–6,000 K (compared to about 6,000 K for the Sun) – but still very bright or luminous. The reason for this is their size; they are 10–100 times larger than the Sun but are 100–1,000 times as luminous. Supergiants are up to 1,000 times the size of the Sun and are thousands of times as bright. The remaining 9% of stars are white dwarfs. These are Earth-sized and are the remains after a star has burned out. They are very dim despite high surface temperatures. Finally, there are brown dwarfs, which are a cross between failed stars and superplanets, but these are not part of a traditional H-R classification or chart.

The creation of the H-R diagram allowed an ordering of stars not previously possible. Moreover, as astronomers later in the twentieth century would learn, the chart also revealed or suggested an evolutionary path for stars. They would form, settle into certain patterns within the main line sequence, and then reach an end by way of a red giant or a white dwarf. This classification and knowledge of stellar history was only possible because of spectroscopy and the knowledge it produced. What was learned was that star types were distributed across the universe, including in M31.

Stellar Distance

The brightness or luminosity of stars is based on surface area, temperature, and distance. All things being equal, the closer a star is to Earth the brighter it will appear. Yet how does one determine stellar distance so as to be able to calculate real magnitude or luminosity?

Calculating cosmological distance remained a problem into the late nineteenth century. From the ancient Greeks to the beginning of the twentieth century determining distance was a problem. For many years it was assumed by Ptolemy and others in the west that all stars were equidistant from Earth. It was also assumed that cosmological distances were much shorter than now calculated. The entire universe was only a few thousand or tens of thousands of light-years. Use of stellar parallax to compute distances to stars proved inexact because stars were so far away their apparent shift was practically negligible. Needed was some tool to calculate distance. This is where luminosity potentially comes in.

Imagine a star shining in space. It is a globe emitting light in all directions. As distance increases, the brightness appears to dim. This relationship can be described mathematically as follows:

$$b = L / 4\pi d^2$$

where b = brightness of a star's light as measured in watts per square meter, L = luminosity in watts, d = distance to a star in meters, and $\pi = 3.1415927$.

Brightness varies inversely squared the further away from the star the viewer is. A viewer twice the distance from a star compared to someone else will see the star one-fourth as bright ($1/2^2$), three times the distance will be one-ninth ($1/3^2$) as bright, and so forth. But to be able to calculate luminosity, one needs to know distance, and this was the problem, at least in the nineteenth century, given that the use of parallax to determine it was inexact, at best. Spectroscopic parallax could be used to determine distance. Once a spectral analysis of a star was done and it was classified, one could arrive at its luminosity and estimate it size. When this was completed then distance could be calculated. But this tool, too, was inexact. The H-R classification is crude; temperatures can vary even within a classification, thereby leaving open many questions about luminosity. A better test at the end of the nineteenth and beginning of the twentieth century was needed to ascertain stellar distance.

Three phenomena came together to help with this calculation. These were variable stars, the Andromeda Nebula, and Henrietta Leavitt.

Variable Stars

Look at the stars on any given night. They are fine pinpoints of light that appear to twinkle. The twinkle that we see is due to our atmosphere. Light from the distant stars reaches our eyes by passing through the atmosphere, with a combination of it and gravity bending light enough to make it appear that they are twinkling. Yet despite this twinkle, the magnitude or brightness of stars remains constant for the vast majority of stars.

There are a few objects in the sky, however, that change magnitudes over time. The planets, for example, demonstrate a significant range or change in magnitudes. All of the planets change in brightness from our standpoint as they orbit the Sun. This is due to their distance from Earth, from the Sun, and the angle at which the Sun's rays strike the planet and are then reflected back into space and to Earth.

The apparent brightness of planets is easy to explain. In part this is because they are not bodies that produce their own light; instead

FIGURE 5.6 Mira at its brightest magnitude of 2.0.

they reflect light from the Sun. But some stars do change magnitudes. Consider the case of Mira, a star in the northern constellation of Cetus. Since the 1600s, astronomers noted that over the course of 11 months Mira demonstrated dramatic swings in brightness – ranging from a maximum magnitude nearly approaching 2.0, and then fading to 4.9. The magnitude or brightness of Mira changes so much that at its dimmest it is about 6% compared to its maximum.

Why does Mira change in brightness so much? This was a question that captivated many astronomers throughout the nineteenth century, and it would only be answered in the twentieth. In the twentieth century astronomers would eventually reach several conclusions about Mira, placing it in a class of stars known as variables.

One explanation for stars such as Mira is that they are part of class of aging red giants whose temperature hovers around 3,500 K. These stars seem to eject significant amounts of gas and energy into space, but they do so in somewhat long and irregular cycles. Stars such as Mira are referred to as long-period variables. While

Mira seems regular in its 11-month cycle, not many long-period variables follow this cycle or this regularity. Why aging red giants become variable is not quite understood. Perhaps it is similar to an old fire that flickers as it begins to die out because its fuel is getting spent. At some point the flame seems brighter, perhaps dims, and then picks up against as a reserve of fuel is found to stoke it again.

But there are other stars known as Cepheid variables. These stars are named after those found in the constellation Cepheid, specifically the star delta Cephei. This star was discovered by English astronomer John Goodricke in 1784. He noted that the star changed brightness in a regular cycle from maximum to minimum approximately every 5.4 days. Astronomers think that the outer envelope of gases or the surface of Cepheids expand and contract. The star literally appears to pulsate, with magnitudes or luminosity increasing during contraction. The cause for the pulsation is not completely understood, but the reason for the increased brightness as the surface collapses is that the contraction causes the star to heat up, thereby producing more energy output and brightness.

There is also another class of stars, RR Lyrae variables, named after a star found in the constellation Lyrae. These variables have cycles of barely 1 day. They are generally found in globular clusters. They, too, are unstable, pulsating rapidly, with the resultant change in size yielding fluctuations in temperature. Together, long-period, Cepheids, and RR Lyrae are known as pulsating variables.

Finally, stars can vary in their apparent brightness because they are part of a binary star system. These are stars that are close to one another. One star may rotate around another. When the stars are side by side (at least in our view from Earth) the brightness of the two stars is added together to produce a maximum brightness. When one star appears to come in front of another, the former partially cancels out the latter and the apparent brightness of the two dims. This is an eclipsing binary. An example of this is Algol in Perseus, which has a 69-h cycle of brightening and dimming. One star is Algol A rotating around Algol B, eclipsing the latter, and producing a temporary dimming of magnitude. This is what happens when binary stars do not touch or have contact. But in some cases the two do touch, with one star pulling mass and energy off the other, resulting in one star getting brighter at times. This is what happens with Beta Lyrae.

Pulsating variable stars of the three types noted above are curiosities. They are difficult to fit into the H-R diagram, because we are not sure how to classify them and also in terms of understanding the exact causes of the variability. However these stars are also important because they may tell astronomers something about the evolution and life cycle of these objects. Finally, their existence and cycle of brightening and dimming has become important in solving another astronomical riddle – cosmological distances.

The 1885 Variable Star

The consensus at the beginning of the twentieth century remained that nebulae were not distant spiral galaxies but instead were collections of stars located within the singular island-universe called the Milky Way. This agreement was held in spite of several

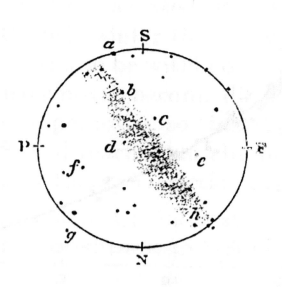

FIGURE 5.7 1886 sketch of the central region of M 31 indicating a new star.

pieces of data that challenged it. For one, Lord Rossi's observations of M31 revealed some spiral shape to the nebula, questioning its nature as either gaseous or perhaps as being a globular star cluster. Yet English astronomer Richard Proctor (1837–1888), best known for one of the first maps produced of Mars in 1867, found a few of these spiral objects in the Milky Way's galactic plane, suggesting to him and others that they were not distant objects but instead part of the galaxy. Yet a more powerful event occurred with the appearance of a supernova in M31 in 1885.

On August 19, 1885 the Irish amateur astronomer Isaac Ward discovered a new bright-red star in the Andromeda Nebula. Subsequently the star was also verified by Estonian Ernst Hartwig. The star brightened to a magnitude of 6, but by early February 1886 the star had faded to 16. What aroused interest and curiosity among astronomers was that previous observations of the nebula had not detected or noticed this star. What astronomers were observing was a supernova.

A supernova is a sudden brightening of a star by many magnitudes. Today astronomers categorize supernovae as Type I or Type II. In general a supernova represents the end of the life of a star. Type II supernova occur with high-mass stars. Essentially what occurs with stars of masses greater than eight or more times that of the Sun is that as they age the core starts to burn hotter and hotter. This occurs because the core is collapsing. The core collapse means an increased temperature. Consistent with Wien's law, the increase in temperature means more energy is emitted from the star, thereby increasing its brightness or luminosity. With a supernova, the temperature due to core collapse reaches hundreds of millions of K.

What occurs with a supernova is that as the core collapses and the temperature increases the former becomes even denser. Moreover, the nuclear reactions taking place in the core produce heavier and heavier metals. At some point the core fuses or creates iron. At this point the density is so great it is almost impossible for any more collapse of the core to occur. So the core essentially bounces back. By that we mean that the reverse of a collapse occurs and suddenly a massive outward-moving shockwave or energy reaches the surface of the star, ejecting it with massive force. This force, equal to hundreds or more times the energy than our Sun

FIGURE 5.8 M1 Crab Nebula.

has emitted in its entire 4.6 billion year life, occurs almost at once. The burst of energy pushes away the outer shell of the star over a short period of time – weeks or months – before it fades away. During this brief period of time a star will suddenly increase in brightness by scores if not hundreds of times. An example of a Type II supernova is M1, the Crab Nebula. It is the result of a 1054 supernova observed by the Chinese.

A Type Ia supernova is the explosion of a white dwarf star at the end of its life. An example of this is the 1572 one observed by Tycho Brahe. NASA's Chandra and Spitzer telescopes later produced images of the supernova.

Supernovae are also listed or named by the year discovered and then by letters following the year, indicating its order or discovery that year. Thus, SN 2012 A would indicate it to be the first supernova detected in that year, SN 2012 B the second, and so forth.

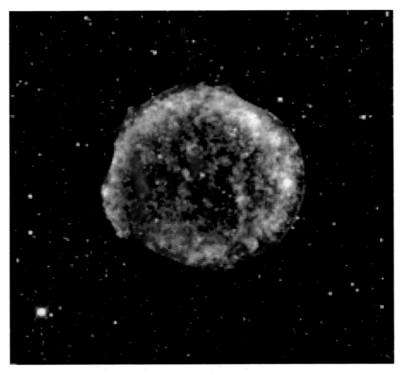

FIGURE 5.9 Tycho Brahe's supernova.

The supernova S Andromedae (SN 1885 A) challenged astronomers. In a brief period of time this supernova brightened and equaled one-tenth of the entire luminosity of M 31. In 1885 astronomers lacked a theory of physics to explain such phenomena of one star brightening and equaling that of hundreds of thousands of others, so they had to resort to other explanations. One theory was that the star flared up as it passed through the nebula. Howard Shapley, in the Great Debate of the 1920s, would point to S Andromedae as evidence against distant galaxies and as supporting the claim that M31 must lie within the Milky Way. His reasoning? Only a star so close could appear so bright. This was the general consensus of the astronomical community during the late nineteenth and early twentieth centuries. S Andromedae was not technically a variable star, but it was still an object whose brightness changed, posing problems for astronomy. But lacking any other theory, this variance in luminosity, a sudden brightening and then fading away, could only be the result of a process taking place in an object not far from Earth and well within the domain of the Milky Way. Yet while its variance might be accounted for by contending that it was passing in and out of stellar gas, it did not explain other variables or address their distances from Earth.

Leavitt, Andromeda, and the Measuring of the Universe

Henrietta Leavitt (1868–1921) is an important figure in the emergence of modern astrophysics. Her importance has only recently been recognized in books and other scholarship. Along with Caroline Herschel, she is one of the unsung female pioneers of modern astronomy. Historically, she is famous for the discovery of a relationship among luminosity, period, and distance with Cepheid variables, which led to developing an important tool for computing stellar distances. It is the work of Leavitt that provided the first meaningful alternative to stellar parallax as a way of ascertaining the distance to stars.

FIGURE 5.10 Henrietta Leavitt.

Leavitt worked at Harvard Observatory. She graduated from Radcliffe College, unable to attend Harvard because of her gender. Not until her senior year in college did she take a class in astronomy. Upon graduation she was unable to find work as an astronomer, but in 1893 she was hired by the Harvard Observatory to do menial work counting and cataloging the brightness of stars from the many photographic plates that the school had. At the time she was hired women could not operate or use the telescopes, and that held true through the early part of the twentieth century. Leavitt was one of several women who worked for Edward Charles Pickering at Harvard.

Levitt was assigned the task of observing stars in the plates made of the Magellanic Clouds. The big and small Magellanic Clouds are two southern hemisphere nebulae named after Portuguese explorer Ferdinand Magellan (1480–1521), who rediscovered them during his circumnavigation of the world around 1520. Emphasis here is upon rediscovery because the clouds had been seen by Arabic and other astronomers before Magellan, but European astronomers had not noticed them because they are located below the horizon. These clouds are best viewed at low northern latitudes and south of the equator. Today astronomers know the Magellanic Clouds to be two irregularly shaped galaxies in close proximity to the Milky Way.

In studying the Magellanic Clouds, Leavitt's attention turned to Cepheid variables. While examining them she noted a relationship between their changes in luminosity and their periods between maximum and minimum brightness. Yet in examining them across the entire sky no apparent pattern between period and luminosity seemed to exist. However, Leavitt turned her attention to the Cepheids in the small Magellanic Cloud, where she presumed that all of the stars were approximately the same distance from Earth. If some of these stars were brighter than others one could conclude them to be more luminous than others.

In charting the brightness against the period, Leavitt found a relation – the brighter the variable the longer the period. She reported these results in 1912. As she stated in that report: "[A] remarkable relation between the brightness of these variables and the length of these periods will be noticed".

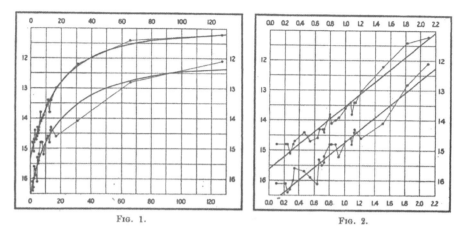

FIGURE 5.11 Henrietta Leavitt's graphs of brightness and periods of Cepheid variables in the small magellanic cloud.

When she re-plotted the relationship a definite linear relationship was revealed. She concluded that a star, for example, nine times brighter than another must be three times farther away. Her discovery, announced in 1912 in "Periods of 25 Variable Stars in the Small Magellanic Cloud," helped establish the luminosity-distance formula for determining the proximity of two Cepheids from one another.

Simply stated, what Leavitt discovered is that the longer the period between the maximum and minimum, the greater its luminosity. Once one knows the average luminosity for a Cepheid one can then calculate its distance using a variation of the formula $b = L/4 \pi d^2$ that was introduced earlier in this chapter. All that had to happen was to establish a true distance to Earth for one Cepheid, and that one could serve as a benchmark for others. The work of Harlow Shapely and others using parallax did just that, setting up Leavitt's discovery as a means to calculating the distance to other Cepheids, and eventually to nebulae such as M31. Leavitt's discovery would prove to be critical for Edwin Hubble in the 1920s, who noticed a Cepheid variable in the Andromeda Nebula and eventually used it to calculate distance and thereby resolve questions about size of the universe and whether M31 was part of the Milky Way or a separate galaxy.

The Nature of Light and the Doppler Effect

Leavitt's discovery of period luminosity for Cepheids was an important benchmark for measuring stellar distance. It looked at how light and energy moved across time, revealing how distance or space and time were related. Yet her research was not the final word on stellar measurements. Other variable stars have also be used as benchmarks, but yet another property of light would prove critical here – the Doppler effect.

The simplest way to explain the Doppler effect is with noises – the sound of a siren. As a fire engine or ambulance approaches the noise seems to get louder, while as the siren races past and gets further away it decreases. The critical factor here is motion. According to Czech mathematician Christian Doppler (1803–1853), the observed wavelength of light is influenced by motion. As an object moves toward us the wavelength it emits appears shorter. In the case of sound, this means the pitch increases as the wavelength shortens. Conversely, as an object moves away the wavelength stretches out, the distance from one peak to another of a wavelength increases, and the sound

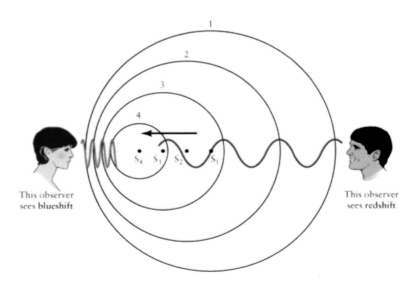

FIGURE 5.12 The Doppler effect depicting redshift and blueshifts.

decreases. What is happening is that in the former the wavelength is decreasing and the frequency increasing, while in the latter the reverse is occurring. This is why an approaching and receding siren appears to change its volume even though the actual siren has not increased or decreased its real volume.

This principle can similarly be applied to light according to Christian Doppler and subsequent astronomers. If in fact objects in space move – such as the planets – then one can determine if they are moving toward or away from Earth by examining how light from then is shifted. In general, objects moving toward Earth will demonstrate shorter wavelengths and their spectral lines will appear shifted to the blue range. Objects receding from Earth will produce longer wavelengths and appear to shift to the red range of the spectrum. This effect is called the Doppler shift. Astronomers call these shifts blueshifts and redshifts. Using spectral analysis to determine if an object is approaching or receding is interesting, but more profoundly, one can then actually calculate how quickly it is moving. This is the velocity of the object. This velocity (v) can be calculated with the Doppler shift equation.

$$\frac{\Delta \lambda}{\Delta_0} = \frac{v}{c}$$

where $\Delta \lambda$ = wavelength shift of the observed object, Δ_0 = wavelength of observed object if not moving, V = velocity of the object as measured along the line of sight, and C = speed of light, or 3.0×10^5 km/s (approximately 186, 282 miles per second).

There are several reasons why calculating the redshift or blueshift and the corresponding velocity are important. First, it simply provides data or information regarding the speed at which objects are moving in space. But understanding the velocity would become important for additional reasons later in the twentieth century, when the Doppler shift was examined in terms of not just planets but also distant stars and galaxies. Many of these objects revealed significant velocities of hundreds if not thousands of meters per second. These velocities questioned the assumptions about how small the universe was.

Second, Edwin Hubble in the 1920s showed that most objects have redshifts and are receding from Earth. One of the most notable exceptions is the Andromeda Galaxy, which is moving closer to Earth and at some distant time it and the Milky Way will collide and merge. If most objects have redshifts and are receding from Earth, this suggests a universe that is expanding. Moreover, Hubble would construct a formula to connect velocity to the distance of a galaxy (along with using what has come to be called the Hubble constant). Calculating galactic redshifts created a new way to measure distances in the universe and ultimately to estimate its size and age. Thus, spectral analysis eventually would be a new astronomical tool that would advance astronomical knowledge across numerous fronts beyond telling us something about the chemical composition of cosmological objects. This analysis would produce a more firm basis on which to make claims about the size of the universe and whether objects seen in the sky were connected to the Milky Way.

The Speed of Light

The Doppler shift equation assumes a speed of light of 3.0×10^5 km/s. How do astronomers know that this is in fact the actual speed? At one time astronomers assumed light moved instantaneously, meaning that it moved at an infinite speed or velocity and arrived at the instant it left the source. In most cases light does appear to move instantaneously. For short distances a speed of 3.0×10^5 km/s is effectively instantaneous. But if this claim were generally true, a distant star would have its light reach Earth as soon as it began shining. Then one is confronted with Obler's paradox: Why is the night sky not completely bright from the light of the distant stars? Should they not all reach Earth at the same time and therefore provide enough illumination to make the night sky as bright as the one brightened by the Sun?

In 1676 Danish astronomer Ole Rømer (1644–1710), after studying Jupiter's moons, reached the conclusion that light traveled at a finite speed. But it was James Maxwell in the 1860s, in his work on electricity and magnetism, who was instrumental in arriving at the conclusion regarding the speed of light. He concluded that light and all forms of electromagnetic radiation move through the universe at a uniform speed of light. Yet that did not end the issue. Physicists assumed that light as a wave had to move through a medium. They reached this conclusion by indicating that waves in water needed a medium and the same was true for sound. Thus, the postulation was that the universe must be composed of some type of "luminiferous aether" that allowed for the movement of light through it. But the problem was how to test this theory.

Perhaps one of the most famous experiments ever performed in the history of physics took place in 1887 in Cleveland (Case Western Reserve University) by Albert Michelson (1852–1931) and Edward Morley (1838–1923). They understood that light could travel through a vacuum. That must mean a vacuum such as outer space contains the aether. But how to test this theory? Michelson and Morley constructed an interferometer to do this. They used the interferometer to split a beam of white light off a half-silvered mirror. The split light would be bent at right angles and then reflect off of small mirrors back to the splitter. The argument here was that if an aether existed, light would take longer to move if traveling with it than if it moved perpendicular to it. In this test, one should be able to detect a slight delay or difference in time of one of the beams when they recombined. The experiment paralleled what they thought would happen to light as Earth moved around the Sun. If it were moving with the aether winds it should slow down light traveling against it, and speed up when assisted by the wind. The assumption was that their experiment would test this.

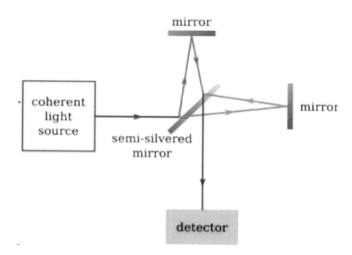

FIGURE 5.13 The Michelson-Morley experiment.

The experiment failed. Yet this is the most famous failure in the history of physics! What they found was no difference in the velocities of light, suggesting that the aether did not exist.

However, this failure was important. The Michelson-Morley experiment, along with the work of James Maxwell, influenced German-born Albert Einstein (1879–1955).

In 1905 Einstein proposed his special theory of relativity. In its most simple form, the special theory of relativity states that one's experience of the world or the universe is the same for all constant velocities. What does this mean? If you are traveling in space along a straight line – regardless of your velocity – you will perceive or experience the laws of physics the same as anyone else moving at a constant velocity (or not moving at all), regardless of the direction they are moving. This is the first principle of the special theory. The second principle is that one will always measure the speed of light to be the same, regardless of how fast you move. Standing still or moving 1,000 miles or km per second, it does not matter – the speed of light will still be 3.0×10^5 km/s. This was opposite of what Isaac Newton had asserted. He contended that if one were stationary and a light beam were shot at a person, that person would measure the speed of light differently from a person moving toward or away from the light.

How does all of this get to the point of relativity? Assume an Earthling observes Mars moving at a different velocity than Earth is moving. (Assume also that you are moving in a straight line.) The perception of speed that the Earthling and the Martian will have is different, or relative. The two will experience space and time differently. The implications of this relativity are many. One is that if an object moves past you and then starts moving more quickly, the perception is that the distance of motion will be shorter. Conversely, if a clock races by you in a high-speed jet it will pass off or record time more rapidly than a clock at rest on the ground. The former experience is referred to as length or distance contraction, the latter is time dilation.

Time dilation associated with the special theory of relativity has become the basis of time-travel believers. If one could travel fast enough – beyond the speed of light – one could actually go back in time! Of course this assumes one can travel at velocities quicker than 3.0×10^5 km/s. Einstein contended this was not possible. This is the case for Einstein because even if one were traveling near the speed of light his theory of relativity states that we would still be experiencing light to be moving the same even if taking a walk. One could not experience light arriving before it has left its source while at the same time experiencing it as moving at the constant 3.0×10^5 km/s.

The special theory of relativity also is the basis of perhaps the most famous equation in western history – $E = mc^2$. Here, energy is equal to mass of a particle and c is the speed of light. The formula states that a particle or object has energy embedded within it, the latter of which is released during nuclear reactions. This relationship is predicted by the special theory of relativity because mass and energy are related or relative to the speed of light.

Finally, the theory of special relativity is important because it suggested that there was no absolute point from which to experience the universe. There was no one fixed central point to the universe from which one could state that all motion or time was held constant. One was always relatively in motion compared to other objects. The implications of this assertion are tremendous, amounting to a second Copernican revolution. If the first Copernican revolution was to cast away Earth from the center of a geocentric universe into a heliocentric one, then Einstein effected a second revolution in demonstrating there actually was no center to the universe at all! Humans were displaced into a centerless universe that was experienced in different ways across the cosmos. There was no way to say which position or perception was privileged – all were equally valid.

Conclusion

By the beginning of the twentieth century long-held views about the universe and human nature had been shattered as a result of modern astronomy. Spectral analysis suggested new ways to think about the potential size and composition of the universe. It was now possible to begin the process of measuring stellar objects, to determine their chemical composition, and to make classifications for stars. It was also possible to make statements about the movements of objects in and through space and time. Finally, new theories about light and relativity rendered previous assumptions about the universe in doubt. If there was no center, were humans part of a Great Chain of Being? Perhaps such a chain did not exist.

The science of astronomy that emerged by the early twentieth century also set the stage for challenging the critical assumptions of the nineteenth century – that the cosmos was composed on one small finite galaxy. At the center of that transformation would be the Andromeda Nebula.

6. The Andromeda Nebula and the Great Island-Universe Debate

The astronomical consensus in 1900 was that there were no island-universes beyond the Milky Way. Earth stood within a single Solar System, within a single galaxy that made up the universe. The extent of the universe was the Milky Way. Even though hundreds of years had passed since Copernicus and, more importantly, new technologies had transformed astronomical science in the nineteenth century, the belief in a closed, finite universe composed of one galaxy persisted. The universe of Copernicus and Newton was more or less the one that existed in 1900.

This wall of consensus persisted despite cracks in its edifice. These cracks were a result of new technologies that made it possible to reach conclusions and speculate more openly about both the nature of the Milky Way and the Andromeda Nebula. These conclusions, about their similarities and contrasts in terms of shape and composition, challenged the 1900 consensus.

The Nature of Scientific Progress

The scientific method is a modern approach to producing knowledge. It is premised upon the gathering and testing of empirical data, preferably in a controlled setting, in order to build theories and claims about the world or universe around us. Although not all forms of scientific inquiry allow for replication or reproduction of a full-blown system of experimentation in order to test and gather information, the core premises of the scientific method revolve around the use of empirical data, testing of claims or hypotheses, and the building of some theories about how things are supposed to work. Thus, in astronomy, the scientific method during the nineteenth century led to the discovery of Neptune based on predictions about the orbit of Uranus. The scientific method also helped in the gathering of information about stars – via spectroscopy – to make claims about their chemical composition and eventually to create the H-R diagram to classify them. The scientific method also allowed for Henrietta Leavitt eventually to discover a pattern in period and luminosity among Cepheid variables that made possible a new way to track the distance to stars.

The nineteenth century thus appeared to be a classic model regarding how normal scientific knowledge progressed. This model emphasized that scientific knowledge was much like a brick wall. These walls are built, so to speak, brick by brick. Each layer of the wall is built on a previous layer. Science is often described this way. New knowledge and discoveries are built upon the work previously undertaken by earlier scientists. Capturing this sentiment was a statement made by Isaac Newton who declared: "If I have seen further it is by standing on the shoulders of Giants." Newton claimed to be building upon the works and discoveries of others when he formulated his theories of gravity and other assertions about nature.

If this model of normal scientific discovery is correct, a linear progression in knowledge in the west from the ancient Greeks to the present can be assumed. Anaximander and Anaximenes provided the foundation upon which Ptolemy built. In turn, Copernicus built upon him and Galileo, Kepler upon Copernicus, Newton upon Kepler, and so forth upon to 1900 and then to the present.

Although there is much true to this depiction of science and how astronomical knowledge advances, there are also problems with this model. Think about how the way the universe was contemplated under the models proposed by Ptolemy, with his geocentric model of the world versus Copernicus with a heliocentric depiction.

If one assumes Earth is at the center then all celestial objects move in circles and epicycles, with planets occasionally literally moving backwards. All these are assumptions now known to be false. But for Ptolemy and his followers, astronomy assumed certain basic facts based upon a theory of the universe, and all efforts to perfect their models of the sky and movement had to conform to both. Their view of the universe defined how they looked at it, and it defined what would be considered to be relevant facts. Claims or observations that contrasted with their assumptions had to be bent or made to conform to their assumptions about the universe.

But Copernicus thought about the universe in a different way. He found that by assuming Earth was at the center of the universe it was difficult to construct a model of the Solar System that successfully predicted the orbit of the planets. There were problems that made the geocentric model ever more complex. There were more epicycles, more assumptions that planets could stop on a dime and reverse direction, and simply more movements that seemed odd for large bodies to perform. It was this difficulty in making the geocentric system work that challenged Copernicus to make an entirely new assumption – that the Sun and not Earth was at the center of the universe. In making this new assumption suddenly Copernicus could better account for planetary movement. Under his model not only was he able to make better predictions about celestial movement, but several "facts" assumed to be true under the Ptolemaic model disappeared. They were no longer facts. Specifically, the concept of epicycles disappeared, as did ideas that planets stopped and reversed motion. All of these movements could be accounted for simply by assuming something different.

The shift from a geocentric to a heliocentric model was the Copernican Revolution. It was a revolution in how to depict the cosmos, but also a revolution in science. It was a rejection of one set of ideas for another. In effect, the scientific discovery or breakthrough that Copernicus produced was not premised upon a linear progression of the building of previous facts upon another. It was not just adding another brick to the wall of knowledge. In many ways, it was a smashing down of the old ways and building up of a new one. Perhaps some of the old bricks of knowledge were used, but it was a new wall that was built.

Scientific knowledge often is the product then of both linear progress and revolutions. There is some building upon past facts and observations but there is also a redefinition of what is considered factual. Sense impressions tell us Earth stands still and the cosmos revolves around it. Reality is Earth rotates. Once it was believed the planets moved in perfect circles, then Kepler demonstrated it was ellipses. Some thought objects fell because they all sought to go to the center of the universe, which was Earth. Newton said it was gravity. Each of these shifts occurred as a result of what one could call a scientific revolution.

But why is the concept of a scientific revolution important? Nineteenth-century astronomy experienced a wealth of events that transformed it. New technologies such as photography and spectroscopy made possible new ways of looking at the cosmos. By the early twentieth century the nineteenth century had discovered a new planet, unlocked many mysteries surrounding light, classified stars, developed a technique for improved measurement of stellar distances, and even proposed a special theory of relativity. All of these events definitely provided new facts and were part of a linear progression of scientific knowledge that effected a revolution in astronomy. But what still had not changed was the belief in a universe composed of a singular Milky Way Galaxy of finite dimensions. The Andromeda Nebula was considered part of the Milky Way, not too distant from Earth. Yet a rethinking was about to occur. New facts, assumptions, and a rethinking of the cosmos led to distinct ideas about both what M31 and the Milky Way were.

The Discovery of the Milky Way

Look up at the sky on any clear night. Crossing the night sky overhead in the Northern Hemisphere one sees the Milky Way. Dating back thousands of years ancient humans looked to the sky and saw the same sight any one of us can see on a clear night. But they probably saw it clearer, without the dim of pollution, city lights, and other obstructions that prevent most of us from seeing it crisply. Like us, they no doubt wondered what this thin haze in the sky was. For the ancient cultures, different answers were reached.

The Milky Way's name comes from the ancient Greek galaktos (milky) kuklos (circle). It was the spilled breast milk from the goddess Hera, produced when Hercules suckled too vigorously. Pythagoras thought it was the track or path of the

FIGURE 6.1 The Milky Way.

Sun that had since been burned. Anaxagoras thought it to be the light from stars obscured by Earth's shadow. For Columbian Indians, the Milky Way was a communication channel or path between Earth and the gods. The Mayans, too, saw it similarly. The Hopi and Navajo depict it in their art. The Pawnee saw the Milky Way as the path of the dead, the tail of the spirits. The Chinese and the Indians saw the Milky Way connected to Earthly rivers, including the Ganges, with the latter born from this heavenly river. Overall, both myth and observation covered and described the Milky Way.

Until the invention of the telescope, the Milky Way was thought to consist only of the stars visible to the naked eye. When Galileo turned his telescope to the sky and the Milky Way in 1610, he saw many more stars than before.

Galileo's drawings on the Pleiades, or M45; the Orion Nebula, M42; and Praesepe or the Beehive Cluster, M44, revealed more detail and stars than apparent to the naked eye. In observing these

FIGURE 6.2 Galileo's drawings of the Pleiades.

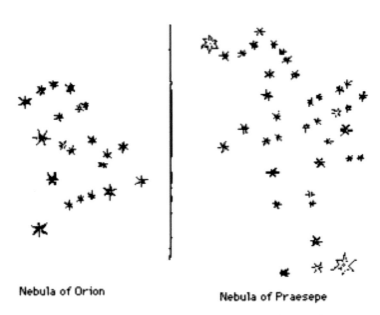

FIGURE 6.3 Galileo's drawings of Orion and the Beehive Cluster.

objects as well as the rest of the sky, including the Milky Way, Galileo concluded that the latter was composed of stars. Suddenly the ancient sky became richer, composed of far more stars than previously thought. His conclusion, as well as his speculation that these new stars were further away from Earth than the others that

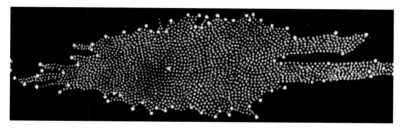

FIGURE 6.4 Herschel's depiction of the Milky Way galaxy.

could be seen without the telescope, led to further questioning about the cosmology and theories of astronomy that were held by Ptolemy and then endorsed by the Catholic Church.

As early as 1750 English clockmaker Thomas Wright suggested that the Milky Way was a large rotating disk. Yet the modern construction or understanding of the Milky Way begins to emerge with William Herschel in 1785. Herschel's diagram of the Milky Way was described by him as the "construction of the heavens". He depicted the Milky Way as a somewhat flat disk composed of millions of stars. It had a fat middle with arms or appendages of stars extending off from the center. Earth and the Sun were seen as being near the center of the Milky Way. How did he reach all of these conclusions? He used statistics. Specifically, he divided the sky up into 683 regions. He then counted the number of stars in each region. He then assumed that stars should be uniformly distributed across the sky. Yet they were not. This led him to assume that the region with the greatest density of stars had to be the center, those with less the periphery. Herschel also agreed with others such as Kant that because we were inside the Milky Way, our impression or view of it is skewed to make it look somewhat flat. Thus, the map or drawing he produced looked as shown here.

Notably significant for Herschel was the observation that the Milky Way was a collection of stars and that he was observing it from within it. More importantly, what emerged for Herschel was recognition that the Milky Way was a galaxy – a collection of stars. He considered the possibility that it was one of many other galaxies floating or adrift in space, but had no way to prove that.

Recognition that the Milky Way was a galaxy led to the nineteenth-century debate over whether it was the singular island-universe in the cosmos or instead one of many others. Some such as Immanuel Kant had argued for viewing it as one of many galaxies, but orthodoxy and the astronomical consensus rested with the Milky Way being the sole galaxy, with other phenomena, such as the Andromeda Nebula, somehow connected to it.

The nineteenth and early twentieth centuries saw significant advancement in knowledge and conceptualization about the Milky Way. Observations of other nebulae, including that of Andromeda by Lord Rosse (Richard Proctor) in the 1840s, of M51, M99, and then the 1871 drawing of M31 all revealed spiral shapes. Given their shapes, Rosse speculated that they might be island-universes similar to the Milky Way (or vice versa). Photographs of M31 by Isaac Roberts in the late 1880s and early 1890s revealed a definite shape. All these observations raised speculation that if they were spiral shaped then perhaps the Milky Way was, too. Moreover, Kant and Laplace in the late eighteenth and nineteenth centuries speculated that the condensation and rotation of nebula into spirals produced solar systems. Thus the Kant-Laplace hypothesis provided a basis for assuming that perhaps even larger entities such as the Milky Way could also be a spiral.

There were others also speculating that the Milky Way might be a spiral. Wilhelm Struve, also doing a statistical analysis of the sky, reached a conclusion similar to Herschel, saying that the Milky Way had a central plane where stars were distributed from a dense core, tapering off the further one traveled from the plane. Princeton astronomer Stephen Alexander published in 1852 a paper entitled "The Milky Way – a Spiral". Richard Proctor, an English astronomer (1837–1888), developed drawings in 1869 that depicted the Milky Way as a ring. Later in his career he suggested that perhaps it rotated. In 1900 Dutch astronomer Cornelius Easton (1864–1929) offered a sketch of the Milky Way that, too, envisioned it as a spiral. Jacobus Kapteyn, a Dutch astronomer (1851–1922), came to similar conclusions as Herschel in the early twentieth century. He concluded that the Sun was at the center of the Milky Way, in a galaxy of 55,000 light-years in diameter. He reached the latter conclusion by examining the brightness and motion of stars. His measurements were off, as later astronomers concluded, because he was unaware of interstellar gases dimming them. Kapteyn's work is significant also

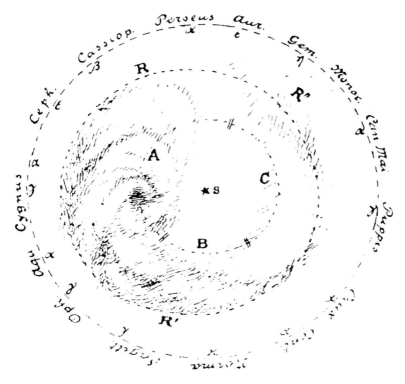

FIGURE 6.5 Cornelius Easton's sketch of the Milky Way.

in seeking to determine the dimensions of the Milky Way, and it also sought out measurements of distances to other nebulae or star clusters. Efforts to obtain these measurements and distances led to significant controversies regarding whether they could in fact be part of the Milky Way.

German astronomer Otto Boeddicker (1853–1937) in the 1880s and 1890s produced a series of drawings of the Milky Way that also lent support to its spherical nature. Overall, during the second half of the nineteenth and into the early part of the twentieth century observations of the Milky Way and other nebulae gave rise to at least three possible conclusions. The first was that the shape of the Milky Way was spherical, with a dense central cluster of stars tapering off toward the edge. For the second, the Sun was at the center of the galaxy. Third, it was thought that many of the nebulae were rotating and that perhaps, too, the Milky Way exhibited a similar movement.

Literally hundreds of articles published in astronomical journals seemed to reach these conclusions. Now we know that the Sun is not the center of the Solar System, and in fact, it is many light-years from the center. But factors such as viewing the galaxy from the inside, the presence of stellar dust, and perhaps even Einstein's concept of specific relativity led observers into being tricked that Earth as an observational point was a central fixed station. There is no question that the other two propositions about the spiral nature of the galaxy and that galaxies rotated were correct. By the early twentieth century, as Charles A. Young, a Princeton astronomer proclaimed in his prominent 1902 textbook *Manual of Astronomy*, the received wisdom and consensus was that the Milky Way was a spiral.

Harlow Shapley of the Harvard Observatory (1885–1972) used RR Lyrae stars in globular clusters to map the halo of the Milky Way in the early part of the twentieth century. As a result of his mapping of the halo and globular clusters, he came to the conclusion that the Sun was not in the center of the Milky Way. He reached this conclusion by recognizing that they were distributed far away from the dense region of the Milky Way in Sagittarius. Based on this, he concluded that the region in Sagittarius was the center of the galaxy, with Earth and the Sun some distance from it and the globular clusters. Thus, we were neither at the center nor edge of the Milky Way, but somewhat off the center and toward an edge. Shapley's mapping also produced distance measurements. If his maps were correct some of the clusters were further away than thought, requiring the Milky Way to be significantly larger than best estimates had suggested. This was because it was assumed that the clusters were part of our galaxy, and if they were so distant, the galaxy had to be much larger than thought.

Sketches and photographs were not the only way to study the Milky Way. Spectroscopic analysis of it was also taking place. J. E. Gill in 1891, for example, found that the majority of stars he observed in the Milky Way were similar to Sirius or the Sun. Other efforts to produce spectroscopic analysis found as a whole that the Milky Way yielded a continuous spectrum that was white. These studies would prove to be important later when analysis of Andromeda and other similar nebulae were undertaken. They, too, would produce similar spectral results.

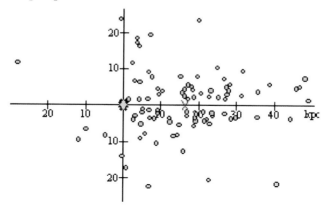

FIGURE 6.6 Shapley graph of globular clusters.

Overall, by the first couple of decades of the twentieth century the Milky Way had been extensively studied. These studies produced significant evidence or basis for the view that it was a spiral galaxy of stars. Yet it was still assumed that there was the single galaxy in the universe.

The Andromeda Nebula

Conclusions about the nature of the Milky Way Galaxy were generated in two ways. One was by directly studying the galaxy itself, the second by examining other phenomena or nebulae in the sky and drawing analogies or comparisons between them. Among the objects that were extensively studied included M31.

Several events contributed significantly to the study of the Andromeda Nebula in the second half of the nineteenth century. The first was the 1871 drawing of it by Lord Rosse, made with a large reflector telescope. It suggested a dense center and a spiral shape. This drawing was followed by an 1888–1890 photograph by Isaac Roberts revealing a clear spiral shape. It was the first photograph that captured what is now the familiar image of M31 that shows its core and the spiral shape to it.

Another event was S Andromeda, the 1885 supernova. Discovered by several astronomers at the same time, it caused great confusion and curiosity. Explaining how a star could suddenly get so bright required understanding stellar structure, type, and star evolution. Studying S Andromedae, as told in Chap. 5, furthered the study and research of stars, eventually facilitating theories about supernovae. But at the time S Andromedae appeared, one conclusion reached to explain it was that M31 could not be that far from the Milky Way, for if it were then the star could not have been so bright. This bright star thus confirmed theories that the Andromeda Nebula had to be connected to the Milky Way.

Another influence on what was being thought at the time was spectral analysis of Andromeda. From the time spectral analysis was being performed in the latter half of the nineteenth century, astronomers naturally looked at M31. What they found was that it had a white continuous spectrum, similar to that found among other nebulae, and also similar to that of the Milky Way. This spectral analysis suggested a range of elements in the nebula, much like that found in the Milky Way. These similarities again seemed to confirm that M31 and the Milky Way were connected.

However, spectral analysis of M31 charting its movements indicated something different. Light as a wave would demonstrate a Doppler effect with its spectrum. If the nebula were receding from the Milky Way it would demonstrate a redshift; if it approached it would yield a blueshift. One individual who sought to measure the shift of M31 was Vesto Melvin Slipher (1875–1969).

V. M. Slipher worked at the Lowell Observatory in Flagstaff, Arizona. Among notable accomplishments, Slipher hired Clyde Tombaugh, the individual who discovered Pluto at that observatory in 1930. But Slipher also did spectroscopic analyses of nebulae. His spectroscopic analyses of nebulae in 1912 at the Lowell Observatory was made possible by the improvement of photography. Together they created solid pictures of the spectral lines. Slipher's spectral work began with M31 as he attempted to discover the red- or blueshifts of nebulae. He found a surprise – M31 had a blueshift, or was approaching Earth, by no less than 300 km/s, whereas other spiral nebulae had redshifts in excess of 1,000 km/s.

Based upon his study of M31, Slipher concluded in 1913 that spiral objects such as Andromeda had higher velocities than did individual stars. He suggested that these higher velocities and motions were indicative of some more pronounced movement in the universe. Slipher's research and questions foreshadowed the discovery of galactic shift or the movement of galaxies in space, prior to the work of Edwin Hubble in the 1920s and 1930s. F. G. Pease in 1917 did even more detailed work calculating the rotation and radial velocity of the central core of Andromeda, computing it at 316 km/s.

But Slipher's computations led to other research that began to question the distances to spiral nebulae such as M31. American astronomer Edward Barnard in 1917 claimed that if the Andromeda Nebula were in fact in motion at Slipher's velocity then perhaps its location in the sky relative to the stars would have changed. Drawing upon extensive observations of it over nearly a 20-year period Barnard concluded that it had not demonstrated any visual displacement or relative motion in the sky. If it were in fact moving as quickly as the studies suggested, some parallax or movement in the sky relative to other stars should have been apparent. Since that had not occurred, the logical or empirical

FIGURE 6.7 Vesto Melvin Slipher.

conclusion of Barnard's work was that M31 had to be further away than previously thought.

Robert Wilson in 1917 reported on the variability of T Andromedae. Like others who followed up on Leavitt's work to ascertain stellar distance through luminosity and brightness, Wilson looked to variables to provide clues. Instead of using the Small Magellanic Cloud as Leavitt had done, he used M31, inspired by the 1885 supernova. Wilson found that the brightness varied with a period longer than expected but consistent with its light curve. Its brightness maxima/minima of 8.45–13.8 visual magnitudes over approximately 132 days indicated a distance greater than would be expected were the nebula within the Milky Way. R. O. Redman and E. G. Shirley in 1921 and J. H. Reynolds in that year also used M31 to study luminosity, finding light curves and variability in the nebula that implied distances greater than one would expect given current assumptions about the Milky Way.

Overall, the research of Slipher and others portended a challenge to the distance and nature of the Andromeda Nebula. What was beginning to occur was the production of a significant amount of research questioning basic assumptions of astronomical research at the time. If the Andromeda Nebula and Milky Way were both spirals, what did that mean? Second, there were other spirals observed, and they, along with M31, were moving at significant velocities. Yet this movement did not produce any apparent parallax or relative motion in the sky. Again, why? Third, the velocities of these objects, especially Andromeda, raised questions about how far away they were in the sky. Fourth, the brightness and luminosity of distant objects, such as S Andromedae, raised questions about stellar distances. Fifth, spectroscopic analysis of Andromeda compared to the Milky Way revealed many parallels. Was this indicative of the two phenomena being one or of something else?

The results of all these observations provided significant anomalies for astronomers. These observations did not readily fit into the prevailing assumptions that the universe was composed of one island galaxy – the Milky Way – and that the size of the universe was rather limited. All this counter evidence or these facts were forcing challenges much in the same way that increased measurement accuracy led Copernicus to challenge the Ptolemaic geocentric model. The time was ripe for a serious questioning of astronomical dogmas; the stage was set for a great debate.

The Andromeda Nebula and the Great Debate

Slipher's discovery of the radial velocity of some spiral nebulae was not unique. Others also found various velocities. However, alone, these velocities did not prove that the nebulae were island-universes. Some contended that the red- and blueshifts were significant evidence of this hypothesis, including American astronomer Heber Curtis (1872–1942).

In the second decade of the twentieth century Curtis worked at the California Lick Observatory, where he undertook nebular photography. Among his objects, several were edge-on. This suggested to him a reason why nebulae were not seen close to the Milky Way's galactic plane; they were obscured by matter in our galaxy. This obscuring by our galaxy and their redshifts led Curtis to speculate that these spirals were distinct island-universes and not something within the Milky Way.

But S Andromedae now reappeared, at least figuratively. At the same time that Curtis was doing his photographic work George Willis Ritchey (1864–1945) in 1917 took photographs of other spirals, finding even fainter nova than the S Andromedae of 1885.

Specifically, he reported in 1917 and 1918 that four faint novae were photographed with a 60-in. reflector. He also reported faint novae in NGC 6946, 2403, M81, and M 101. Regarding the Andromeda Nebula he concluded that it was either very large or very near given what he was able to observe compared to other spirals he was able to see in more detail. Also, based on his study of M31, he reasoned that other spirals probably demonstrated similar characteristics in terms of internal rotation, along with proper motion and the dark rifts he found in it. Ritchey had detected internal rotation in M31 along with M81. Other astronomers also detected similar rotations in spirals.

The varying brightness of these novae could be evidence of different classes of these objects, or perhaps of even different distances

FIGURE 6.8 George Ritchey.

to the nebula. Curtis took evidence of novae in distance spirals as a sign that the spirals were much further away than thought, perhaps even more distant that the confines of our galaxy.

At the time Curtis was doing his work the debate begun by Kant regarding the existence of island-universes reemerged. Hector Macpherson in 1919 began his discussion of the island-universe theory by quoting Agnes Clerke's conclusion that there was a clear consensus that there was no star system that ranked as equal to the Milky Way. Macpherson declared that more reliable measurements of distances of star clusters and spirals, along with spectroscopic and theoretical considerations, were making it increasingly difficult to contend that the Milky Way was the sole galaxy in the universe.

Macpherson cited Slipher's computation of radial velocities of radial nebulae such as M31, and Howard Shapley's research on distance to star clusters, as helping to reopen this debate. Macpherson also referenced the appearance of "temporary stars in Andromeda" as revealing a distance to M31 of 1,000,000 light-years and a diameter of 50,000 light-years as igniting the island-universe theory. If these objects were so distant and large, how could they be part of the Milky Way unless the latter was immensely large?

FIGURE 6.9 Heber Curtis.

Other astronomers also suggested that estimates on the size of the diameter of spirals might be influenced by their distance, with some appearing smaller due to the greater distance from our galaxy than others. In addition, among the spirals observed, not all of them seemed alike, with M31 perhaps being of one type that was similar to the Milky Way.

Although many astronomers engaged in the island-universe debate, the two most vocal or famous combatants were Heber Curtis and Harlow Shapley. Shapley's research on star clusters and their distance led him to argue in favor of the "Big Galaxy" theory, while Curtis viewed evidence supporting the island-universes hypothesis. They extensively argued their points, with Curtis seeing the new data as evidence of multiple island-universes, Shapely contending that the evidence supported viewing the Milky Way as an immense galaxy, containing other spirals including M31. In 1920 a Washington, D. C., meeting between the two, termed the "Great Debate," was arranged for them to argue their rival positions. However, leading up to the debate, both Curtis and Shapely dueled their positions as academics often do – in print.

Shapley engaged the debate on external galaxies by first contesting an assertion of Curtis. Specifically, when Curtis and others

FIGURE 6.10 Howard Shapley.

had observed nebulae they assumed that the novae associated with them were connected. Shapely disagreed with that connection. He also questioned whether spirals associated with the nebulae were connected. The purpose in questioning these connections was that if the novae or spirals associated with the nebulae were empirically distinct, then any redshifts or use of the luminosity-period function to calculate distance and thereby situate the nebula at a distance far outside the Milky Way would be inaccurate. The shifts or distances measured would not be that of the nebulae but of the spirals.

Second, if distant stars in the spirals are connected with the nebulae and they are assumed to be as bright as the stars in the Milky Way, then that would imply two things for M31. First, it would have to be at least one-million light-years distant, and, second, it would have to be at least 50,000 light-years in diameter. Both of these measurements were rejected by Shapley as too large and improbable. Third, Shapley argued that measurements of the internal proper motions of the spirals are inconsistent with the island-universe theory. If the velocity of M101 was 1,000 km/s, then it would only be 32,000 light-years away, making it possible to resolve stars within it. However, since these stars in M101 and M31 were not resolvable these objects could not be island-universes.

Shapley built upon these arguments in his 1919 essay "On the Existence of External Galaxies". He again contended that their high speed was not proof of their being external or island galaxies, but he also further argued that if they were galaxies then individual stars and the center should be able to be resolved, but they had not been. Shapley employed the argument (refuted perhaps earlier by Curtis) that their absence along our galactic plane is proof that they were within the Milky Way. Moreover, recent research seemed to upgrade the size of spirals to be over 300,000 light-years in diameter. This immense size was rejected since proponents of the island-universe theory saw these objects as only supposing to be the size of the Milky Way. Thus, facts driving the size of these galaxies were inconsistent with the claims made about how large they were supposed to be.

Shapley additionally drew upon research by Dutch-American astronomer Adriaan van Maanan (1884–1946), who at the Mount Wilson Observatory had computed the absolute luminosities of distant stars. If his data were correct then the novae in these galaxies should far exceed any known luminosity. Similarly, the galaxies would or should be brighter than expected from the current information we had. So Shapley concluded that the internal motions, radial velocities or redshifts, brightnesses of stars, and the spirals themselves all pointed away from the belief in other island-universes.

Heber Curtis argued for the island-universe theory in two articles or papers. First, in his 1917 "Novae in Spiral Nebulae and the Island-universe Theory," he contended that the brightness of the novae in distant nebulae led to two conclusions. First, the distance associated with the brightnesses of these novae would require the objects in which they were located to be millions of light-years away, far outside of the Milky Way. Second, their diameters would need to be at least 60,000 light-years wide. For Curtis, neither of these dimensions seemed to be possible, as was asserted by Shapley.

In his 1919 article Curtis drew upon many of these and previously made assertions to defend the island-universe theory. He again discussed the location and distribution of spirals near the Milky Way's galactic plane as one piece of evidence. He drew analogies or parallels between spirals such as M31 and the supposed spiral structure of the Milky Way as an indication of the formers' galactic nature. Curtis also noted how the spectra of the nebulae was consistent with that of star clusters, and he argued that the velocities of their objects, as well as their shifts, suggested these objects were distant and that such a distance should not be rejected out of hand. Finally, he stated that if he was correct then it should someday be possible to resolve the stars in the nebulae.

Given the pointed comments made by Shapely and Curtis, one would have expected the 1920 Great Debate to have been intense and heated. It was neither. Shapely provided two claims at the debate. First, he pointed to the absence of nebulae in the galactic plane as proof that they were part of the Milky Way where new stars were birthed. Shapley asserted that the plane existed as a "zone of avoidance." By that he meant, perhaps stars fell into this zone after being created outside of it. His second claim referenced the 1885 Andromeda nova. Since in 1920 nova as understood today did not exist, it was impossible to explain how one star could become as bright as it did if it were part of a distant galaxy. Instead, the brightness could best be explained by locating it closer to us within the Milky Way.

Curtis responded by attacking both of Shapely's arguments. First he assailed the zone of avoidance. He did that by again asserting that the obscuring effect that the Milky Way had in terms of hiding galaxies in the plane. Second, he dismissed the brightness of the 1885 nova as abnormal; most novae in Andromeda and elsewhere were much fainter and thus consistent with the far distance of island-universes.

The Great Debate Fizzles

Unfortunately, the Great Debate was neither. It failed to resolve the island-universe controversy both because no definitive distance to the nebulae, such as M31, had been determined, and because of the inability to resolve individual stars in these spirals. The latter was critical because unlike the Milky Way, where individual stars could be detected, the latter could not occur with these distant objects. To resolve the debate one needed either to be able to calculate specific distances or resolve stars in the spirals. One also needed a theory that could account for the brightness of novae and the red- and blueshifts detected among spirals. There were many unanswered questions left over from the Great Debate, and they would soon be answered by Edwin Hubble, who drew most notably upon the work of Henrietta Leavitt to resolve the island-universe debate.

7. Edwin Hubble, an Infinite Universe, and the Classification of Galaxies

The Great Debate was inconclusive. It failed to resolve a basic dispute critical to understanding the universe: Were we alone in the Milky Way as the only galaxy in the universe or did others exist? Was, for example, M31, the Andromeda Nebula, something distinct from the Milky Way, perhaps a separate galaxy existing on its own, or was it simply part of ours? To conclude that Andromeda was a unique galaxy also had broader implications, suggesting that perhaps other galaxies, in the millions if not billions or more, existed that were distinct from the Milky Way. Additionally, perhaps many of the other phenomena observed in the sky, too, might be separate from the Milky Way? But how to resolve this debate?

The key was evidence that contested the prevailing assumptions about the universe. These assumptions included the belief that the universe was of a finite size such that no phenomena, such as a star, was that distant. In addition, prevailing wisdom was that the Milky Way itself was of rather modest size in terms of light-years in length, and that the Sun and Earth were located near the center of it.

However, increasingly, new facts and evidence began to question the single galaxy assumption. As seen from the Great Debate, the 1885 supernova raised questions about distance. How could an object become so bright unless the Andromeda Nebula was close to Earth? Conversely, the rarity of bright stars seemed more consistent with objects being very distant. But additionally, the similarity in the shape of the Milky Way and Andromeda raised questions about what the latter really was. Spectroscopic analysis of M31, compared to the Milky Way, demonstrated powerful parallels that again suggested similarities between the two. These parallels could be evidence of the two being one object or two distinct objects but displaying similar spectra. Finally, the redshifts and radial velocities of objects in the sky also raised questions about distance. If objects were truly moving that quickly, why could not their visual movement in the sky be detected? Why was no parallax detected? Close-up objects moving that quickly should be changing their position in the sky in relationship to other objects. None of that was seen. Trying to account for all of this was a problem.

Yet these anomalies could be accounted for much in the same way that pre-Copernicans defended the heliocentric view of the universe. Evidence could be ignored, accommodated to the existing world view, or models made increasingly more complex to make the assumptions work.

Prior to Copernicus, one way to make the heliocentric model work was to add more epicycles to make celestial movements and predictions of these movements more accurate. In the early twentieth century similar techniques were employed as noted above. Bright or luminous objects proved their closeness, while dim objects demonstrated obscurity by clouds or gas. Similar spectra proved all objects were part of the Milky Way, parallel spiral structures either ignored as coincidence, and redshifts (or blueshift for Andromeda) explained to show how the Milky Way itself and all the objects attached to it were rotating together. Somehow, no matter what, evidence could be interpreted to fit the existing model and assumptions.

So what would constitute evidence to refute the prevailing paradigm and assumptions? One possibility would reside in challenging cosmic distances. Specifically, one would first have to make an argument about the size or dimensions of the Milky Way Galaxy. Establishing its size would define not only how big it was in terms of length or dimensions but also would provide a benchmark for estimating how far away other objects were assumed to be if they were part of this galaxy. One also needed some type of reliable tool to estimate distances in the universe. Up until the early twentieth century calculation of cosmic distances was inexact and based on guesses. Although stellar parallax was the most often used tool to estimate distance, its utility was compromised because of the distance of stars. Thus, if a tool to reliably measure distance were found, and that tool pronounced distances to objects such as the Andromeda Nebula that were further than expected – especially further away than the best estimated size of the Milky Way galaxy – then that would offer more proof that the universe was larger than expected and that there were objects unattached or distinct from our galaxy.

Finally, there needed to be some theory that accounted for the different shapes of objects in the universe. The Milky Way was assumed to be a spiral, and there were other objects in the universe that were spirals, but there were also other objects that displayed various other shapes, some of which were more spherical in nature, others more irregular. A theory that could account for these different shapes, and then make an argument that they were galaxies, would provide additional evidence to contest the prevailing paradigm. Overall, what was needed was to provide a theory that offered solid proof about stellar distances and dimensions and that accounted for different shapes to establish these objects as galaxies – and do it in a way that provided for a simple explanation – and one would have a Copernican Revolution when it came to the island-universe theory.

Edwin Hubble, Cepheids, and the Andromeda Galaxy

The first break towards resolving the Great Debate came with the work of Edwin Hubble. Hubble (1889–1953) was an American trained astronomer who grew up in Missouri and studied astronomy and earned his Ph.D. at the University of Chicago after also studying at Oxford University. While at Chicago he worked at the Yerkes Observatory, and his dissertation was titled *Photographic Investigations of Faint Nebulae*. He did his doctoral work after teaching high school for about a year. After earning his doctorate he served in World War I. After the war and in 1919 George Ellery Hale, founder of the Mount Wilson Observatory in California, offered Hubble a position there. Hubble remained there the rest of his life.

Mount Wilson is located in southern California, not far from Los Angeles. It had a history with astronomy even prior to Hubble. In 1889 Harvard University placed a couple of telescopes

FIGURE 7.1 The building housing the Mount Wilson 100-in. Hooker telescope.

FIGURE 7.2 George Hale.

there, but due to often bad weather, this small observatory was quickly closed.

Three years later, in 1892, Harvard president Charles Eliot planned for 40-in. lenses to be delivered to nearly Mount Harvard, but the deal fell through and the already built lenses found their way to Yerkes Observatory in Wisconsin. In 1903 George Hale (1868–1938) visited the Mount Wilson site and decided to move ahead constructing an observatory there.

Hale had taught astronomy at Beloit College in Wisconsin and at the University of Chicago. He is an interesting character in the history of astronomy and telescopes. He pushed for bigger and bigger telescopes to be built, initially involved with the building of the large 40 in. refractor at the Yerkes Observatory in Wisconsin in 1897. He was also behind the building of the 60-in. reflector at Mount Wilson in 1908 and then the 100 in. reflector there completed in 1917.

FIGURE 7.3 Mount Wilson's 60-in. telescope.

FIGURE 7.4 Mount Wilson's 100-in. Hooker telescope.

It would remain the largest telescope in the world until 1948, when the Mount Palomar Observatory, also in California, would open with its 200-in. reflector. Hale was also a moving force in the construction of the Palomar Observatory. Hale's motivation in pushing for larger and larger telescopes was simple – a desire to see further into space, gather new scientific knowledge, and construct tools that could resolve individual stars and many of the astronomical controversies of the time, including the island-universe debate.

During his tenure at Wilson Hubble would benefit from having two of the largest telescopes in the world. He would use both the 60- and 100-in. telescopes in his research. But Hubble especially benefited from improved photographic technology and the increased light-gathering power of the 100-in. diameter telescope at Mount Wilson. The clear atmospheric conditions of Mount Wilson also contributed to the ability of Hubble to use these new astronomical tools and technologies to aid his research. But two other things were important to the work his was doing – the prior conclusions regarding variable stars reached by Henrietta Leavitt and the Andromeda Nebula.

As discussed in Chap. 5, Leavitt's discovery of a relationship between the period and luminosity of Cepheid variables in the Small Magellanic Cloud was a brilliant observation and insight into the distance of stars. If Cepheid variables had the same period then one could conclude that they had similar variability in luminosity. Once calculated this meant one could then determine distance. Her discovery gave astronomers a means of calculating stellar distances far superior to parallax that, since its employment with the ancient Greeks, had yielded highly inexact calculations due to the remote distance of stars and the almost insignificant shift in the sky as a result of the revolution of Earth around the Sun. By employing her period-luminosity relationship it should have been possible to calculate the distance to the Small Magellanic Cloud. But something far more significant could also be done – apply it to stars in the Andromeda Nebula.

The 1885 supernova in the Andromeda Nebula aroused significant curiosity. It appeared out of nowhere, and lacking then a theory to explain it, raised troubling questions for the island-universe debate. As Curtis and Shapely engaged in the Great Debate, they pointed to this star in M31, using it either as evidence of the close proximity of the nebula, or as evidence of an anomaly, with the general faintness of Andromeda as proof of its distance. Moreover, the inability to resolve individual stars further reinforced the idea that this object must be very distant. But what if instead of it being a supernova, one could locate Cepheid variables in the Andromeda Nebula? That would make it possible to calculate distance to the stars and then to the nebula in general. If then one could obtain a good and accurate measurement to M31, and it distance was significantly beyond the assumed dimensions of the Milky Way, that would constitute a dramatic step in establishing that the object was not part of our galaxy. It would demonstrate the objects to be independent of the Milky Way, in a universe far larger than previously assumed. Thus, combine Henrietta Leavitt, Cepheid variables, the Mount Wilson Observatory, and the Andromeda Nebula together and one would be able to challenge much of the astronomical consensus. This is what Edwin Hubble did.

164 The Andromeda Galaxy and the Rise of Modern Astronomy

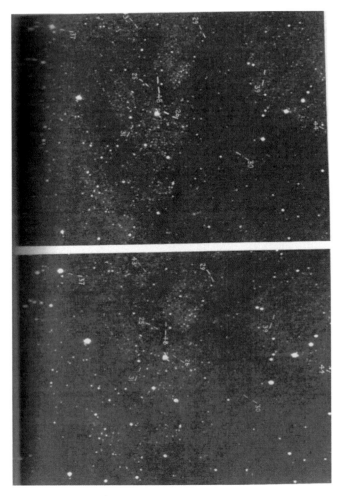

FIGURE 7.5 Cepheid variables in M 31 (Hubble 1982).

Hubble had use of the largest telescope in the world at Mount Wilson. His dissertation addressed astrophotography and nebulae. It was simply natural to turn to the Andromeda Nebula to examine and photograph. At the time he was doing this astronomers still were seeking to resolve individual stars in it, and determine the nature of the spiral shape, the reasons for its blueshift (movement toward Earth), and its similar spectra results.

In October 1923, Hubble was using the 100-in. Mount Wilson telescope to take a 40-min picture of M31. Upon developing he noted a new spot and assumed it to be either a photographic defect or a nova. He retained the photograph. The next night he repeated the photograph and again noticed this spot as well as two others. He annotated the plate with an "N" next to each star he assumed to be a nova. He decided to compare these plates to those which had been archived. Two of the stars did turn out to be nova, but the third was even more of a surprise; it was a Cepheid variable. He crossed out the "N" on the plate and replaced it with a "Var" for variable. This discovery was to be critical to resolving the island-universe controversy because he could use the Leavitt period-luminosity relationship to ascertain the distance to M31.

Hubble found that the period for this Cepheid was 31.415 days, making it more than 7,000 times more luminous than the Sun. What did that mean for the distance to Andromeda? Hubble used this Cepheid and others in the Andromeda Nebula to compute the distance first in his 1926 article on M33 and then in another 1929 article on M31. The latter, entitled "A Spiral Nebula as a Stellar System, M31," is perhaps the most important article ever written about Andromeda and in terms of resolving the island-universe debate. In this article Hubble first reviewed the history of recent observations regarding M31. He noted its spiral shape, its spectrum, its blue shift, and then he described how he had located 40 Cepheid variables in the nebula. Their discovery was made possible by comparing photographic plates of the nebula taken by the 100-in. Wilson telescope. To record the data, he divided Andromeda into different regions, locating stars in each, and then recording their maxima and minima in order to determine their periods.

Once the periods and luminosity for the Andromeda Cepheids were calculated, Hubble used the two to begin the process of calculating their distance. Thus, he drew upon the arguments made by Leavitt in her study of the Small Magellanic Cloud. He used her research, Shapely's calculation of distance in the Small Magellanic Cloud, and his own prior determinations of the distance to M33 – a companion galaxy to M31 – to ascertain the distance to the Andromeda Nebula.

He stated in his 1929 article that the distance of M31 is about 0.1 mag., or 5%, greater than that of M33, and 8.5 times the distance of the Small Magellanic Cloud. Using Shapely's value for the Cloud ($m-M = 1.55$) we find for M31:

$$m - M = 22.2$$
$$\pi = 0.''00000363$$
$$\text{Distance} = 275{,}000 \text{ parsecs}$$
$$= 900{,}000 \text{ light-years}$$

Prior to Hubble, estimates to Andromeda were made using parallax. In 1907, one estimate was that M31 was only 6 pc or a little more than 18 light-years away. In 1922, using estimates of the velocity of Andromeda's rotation, Ernst Öpik (1893–1985), an Estonian astronomer, calculated the distance to be 450 kilo parsecs (KPC), or approximately 1.47 million light-years. This was significantly greater than earlier estimates, and exceeded that of Hubble. With the former estimate of it being a few light-years away, Andromeda could easily be assumed to be well within the realm of the Milky Way. The latter estimate by Öpik raised questions about this assumption. Thus, Hubble's calculation using the period-luminosity relationship added more fuel to the fire that questioned the distance to M31 and the single galaxy theory.

Hubble's calculations of distance using Cepheids in M31 had significantly ended the Great Debate. How so?

The Milky Way was 100,000 light-years in diameter, as was generally accepted by astronomers. According to Hubble, Andromeda at 900,000 light-years distant could not be part of our galaxy unless you then assumed the Milky Way to be at least nine times larger than anyone had assumed or calculated. Thus on distance alone, one had to conclude that Andromeda was far distant and therefore separate from the Milky Way.

If M31 was that far away, how does one account for its brightness? This was one of the issues central to the Great Debate. In that debate, accounting for the brightness of the 1885 supernova led some to speculate that M31 had to be relatively close to Earth and the Milky Way. But here a different conclusion or hypothesis could be offered to account for its brightness. If Andromeda was as far away as Hubble calculated then it had to be exceedingly bright to be seen with the naked eye. Hubble concluded that the brightness had to be a consequence of it being composed of hundreds of millions of stars, much like the Milky Way.

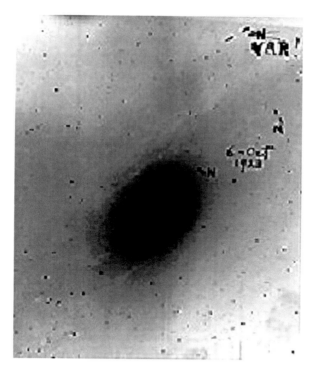

FIGURE 7.6 Edwin Hubble's plate of Andromeda indicating where he crossed out the N (Nova) and wrote Var (variable) instead.

Proof that Andromeda was composed of millions of stars could be determined by luminosity and the calculation of mass in terms of our Sun. Hubble calculated the mass by first locating the central core of M31. Once identified he then indicated that one could use a spectral analysis to calculate a shift. This shift revealed a rotation of the core or nucleus of the spiral that he estimated to be approximately 200 pc in diameter. This rotation was 72 km/s. Given the rotation velocity and diameter he calculated the mass to be $2.4 \times 10^8 \odot$. The symbol $\odot$ represents one solar mass, or the mass of the Sun.

Hubble then assumed, based on his observations of M33, that the entire mass of M31 would be approximately 10–12 times that of the inner region, thereby producing a total estimated mass of $3.5 \times 10^9 \odot$. He then reviewed what astronomers knew about the mass and luminosity of the Sun, using that as a basis for an overall calculation of the mass of Andromeda given its luminosity.

Hubble found that the average density of M31 could not be accounted for (and given its brightness) unless one assumed that it was composed of bright and (pardon the pun) massive objects similar to the Sun. In other words, the only way the mass and the luminosity could be explained was by assuming it was in fact composed of these millions of stars. Therefore, M31 was a spiral object composed of millions or more stars, and not stellar gas. Its mass and luminosity could only be explained if one assumed that it was composed of stars. Moreover, given its distance of 900,000 light-years, its brightness could only be explained if one assumed it was in fact composed of so many shining stars. This is what made it visible despite its significant distance from Earth.

Finally, based on the calculated mass and luminosity, Hubble concluded that M31 had a diameter of 80,000 pc, or approximately 240,000 light-years. So given all this information, here is how Hubble put it together. The Milky Way is about 100,000 light-years in diameter. Leavitt's period-luminosity relationship for Cepheid variables put the distance of M31 at 900,000 light-years away. The spiral shape of Andromeda was similar to that of the Milky Way. The brightness of the nebula could only be explained by fact that it was composed of stars. The mass confirmed this observation. Finally, given its mass and brightness, M31 was no more than 240,000 light-years in diameter. Given its diameter, the diameter of the Milky Way, and the distance between them, Hubble declared that there was only one possible conclusion – the Andromeda Nebula was the Andromeda Galaxy! Andromeda was a distinct galaxy on its own.

Hubble's conclusion was correct, although his use of the period-luminosity relationship and estimates of mass and distance were not accurate given what we know today. The best estimates today are that the Andromeda Galaxy is approximately 2.4 million light-years distant, that its length or diameter is 220,000 light-years, and that its mass was 3.2×10^{11} solar masses. Hubble was close to accurate on Andromeda's diameter, 100× shy on its mass, and a little less than 40% of the now known distance. But even with these inaccuracies, Hubble's calculations were enough to establish his core argument that Andromeda was a galaxy.

The implications of establishing the Andromeda Nebula as an independent galaxy were wide ranging. If Andromeda was a spiral and there were other spirals that had been observed, could they, too, be galaxies? That is exactly what Hubble's discovery and argument implied. It suggested that a host of other galaxies existed. This meant that the prevailing single galaxy theory was incorrect and that instead, Kant had been correct in his assertion that the Milky Way was merely one of many galaxies in the universe.

Finally, Hubble's star, labeled by him and known as V1, is perhaps the most famous star in astronomical history. It remains an astronomical curiosity, still studied and examined by astrophysicists and agencies such as NASA. It was even recorded by the Hubble Telescope. Its Cepheid period, now set at 31.415 days, helped establish and effect a change in a paradigm about how the universe is conceived, similar in scope to what Copernicus did when he proposed a heliocentric universe.

FIGURE 7.7 Hubble's Star in a January 2011 photograph by the Hubble telescope (Courtesy of NASA).

Implications of Hubble's Andromeda Calculations

Examination of the Andromeda Galaxy proved to be critical to a paradigm shift in astrophysics and astronomy. First, it offered proof that Kant and Curtis were right, the island-universe theory was correct. Instead of the universe being composed of the Milky Way alone, there were potentially millions of other galaxies throughout it, composed of billions of stars with perhaps their own solar systems.

Even though some such as Öpik had earlier estimated the distance to M31 as 450,000 pc, few others had agreed with this calculation. Hubble's work on M31 helped to establish some estimates of galactic distance that went beyond previous agreed-upon estimates. If Shapely could object to the island-universe theories on the basis of claiming that the distances were too great to explain their luminosity, the Hubble calculations for Andromeda demonstrated a more expansive universe than previously thought. The distances accounted for the red- and blueshifts that Slipher and others detected when they calculated radial velocities of the galaxies. Moreover, the expanded universe also explained other phenomena. For one, it accounted for why individual stars in these galaxies could not be resolved – they were too distant. Second, the new cosmological distances also explained the apparent lack of parallax among stars – again they were too distant. Third, the failure to detect real movement of stars in the sky, despite the red- and blueshifts of these objects, could also be explained. Even though the objects might be moving a few hundred kilometers per second, this movement could not be detected in the sky because they were so far away.

Yet Hubble's arguments and conclusions posed new problems for astronomers. How does one account for the redshifts observed especially among galaxies? Was it simply a rotation or was something else occurring? In the case of the Andromeda Galaxy it was even more perplexing in the sense that it had a blueshift. In all these cases, how to account for why galaxies appear to be moving away or toward Earth? Eventually, such shifts would have to be reconciled with Einstein's theories on the expansion of the universe.

Another result of Hubble's discovery forced a rethinking of stellar evolution. Shapley had attacked the island-universe theory by arguing that the luminosity of S Andromedae in 1885 was not consistent with the distance needed to support this brightness. However, Hubble computed the light curves of novae, and he, along with others, eventually reached a new conclusion. S Andromedae was not a nova but what we now call a supernova. Neither Shapely nor Curtis had known this in the 1920 Great Debate, but Hubble's work, as well as subsequent research by others, confirmed this new type of stellar phenomena.

Yet another result of Hubble's study of Andromeda was to support Curtis's explanation for why nebulae, or now galaxies, could not be seen near our galactic plane. The plane of the Milky Way obscured detection or observation of these objects. Moreover, if these other objects were not always nebulae but galaxies, their shapes provided clues to what the Milky Way looked like, revealing in some cases a thick central region that in fact could obscure or hide objects on the other side of them if one tried to look through them.

FIGURE 7.8 The galactic plane of the Milky Way Galaxy in infrared light (Picture taken by the Spitzer telescope).

Finally, Hubble would prove to be both right and wrong, as noted above, about the distance, mass, and diameter of M31. Subsequent research would correct and perfect his results. These corrections would also reveal something else that Hubble was both correct and incorrect about – that Andromeda was separate from the Milky Way. Although Hubble was correct that M31 was a distinct galaxy, he was not completely right in asserting that it was independent from the Milky Way. Subsequent research would establish that the Milky Way, the Large and Small Magellanic Clouds, and Andromeda were part of the Local Group, a distinct cluster of galaxies. Astronomers would eventually discover that there are many galactic clusters in the universe, with the Milky Way and Andromeda part of one, and another cluster located in the constellation of Virgo.

Andromeda and the Classification of Galaxies

Hubble's initial work established that M31 was a galaxy. He did the same for M33, the Triangulum Galaxy, and NGC 205, both companions of the Andromeda Galaxy.

All of these galaxies were spirals and therefore easy to argue that they were galaxies. But Hubble's demolishing of the island-universe theory left open at least two additional questions. If galaxies were spirals, why were not all spirals the same? In addition, what about objects or nebulae that were not spirals – were they galaxies?

Consider three popular objects for viewing, the Large Magellanic Cloud, M104, the Sombrero Galaxy, and the Dumbbell Nebula, M27. Looking at the shape of these three objects it would be difficult to determine which were actually galaxies and which are not. None of them display the same spiral shape found in M31, yet astronomers have now concluded that the Large Magellanic Cloud is a galaxy near the Milky Way and that M104, too, is a galaxy. Yet M27 is simply a gaseous nebula and not a galaxy. This is especially curious when one looks at the Large Magellanic Cloud, which does not look like a galaxy at all.

FIGURE 7.9 M33, the Triangulum galaxy.

FIGURE 7.10 NGC 205.

FIGURE 7.11 Large magellanic cloud (NASA photo from the International Space Station).

FIGURE 7.12 The Sombrero Galaxy, M104.

FIGURE 7.13 The Dumbbell Nebula, M27.

Instead, its irregular visual appearance might make it look more like M27 or another nebula. What is the difference between real nebulae and galaxies?

Today astronomers have a variety of ways of distinguishing the two. At the most basic level, the distinction is simple – galaxies are composed mostly of stars, whereas nebulae are not. However, this distinction is not completely accurate, since now astronomers know that many nebulae are at the center of star-forming regions. Specifically, at the center of M42 is the Orion Nebula, which is a massive region of star birth. But back in Hubble's time the lack of data and observations of the two made it difficult to assert that some nebulae were not composed of stars. Perhaps increased magnification would reveal stars in them. Moreover, today astronomers could look to spectra analysis to distinguish the two, but again in Hubble's time there was no good theory to explain any differences in spectra detected.

The problem then was how to distinguish and explain the different-shaped objects in the sky, specifically accounting for the differences between nebulae and galaxies, and then for the differences in the shapes of the various galaxies. If some nebulae were indeed galaxies, one still needed to explain the connection between the two. Were they distinct phenomena, as suggested by Shapley in the Great Debate, or were there star clusters or galaxies embedded within every nebulae, at least as implied by Curtis? Hubble's research offered a solution to this problem with his classification of galaxies and their evolution. He proposed this classification in one of the classic books of astronomy, *The Realm of the Nebulae* (1936). The book was based on a series of lectures he gave at Yale in 1935.

Hubble's work did not end with calculations of the distance to the Andromeda Galaxy. He continued to investigate it and other spiral galaxies, seeking to address and refute many of the arguments levied against the island-universe theory. For example, Hubble took on Shapely's zone of avoidance theory in his survey of nebulae and spirals in the sky. Shapley had used the absence of nebulae in the Milky Way galactic plane as proof that they were connected to our galaxy. Hubble decided to produce a map that pinpointed them. His map demonstrated specifically that this zone followed, according to Hubble, a "continuous irregular belt along the Milky Way with two small and significantly located exterior patches. The general pattern follows the distribution of known obscuring clouds and the zone is presumed to represent analogous phenomena. The irregularity is strong evidence that the obscuration is due largely to isolated clouds rather than to a uniform layer of diffuse material." In effect, what Hubble found was that the zone of avoidance overlapped with the galactic plane of the Milky Way. The reason why no nebulae could be seen in the plane was that the latter was thick with stars, clouds, and gas, thereby obscuring any objects lying beyond it. In terms of the Great Debate, Curtis was correct; the zone of avoidance was wrong.

Yet in the process of doing this survey Hubble photographed hundreds if not thousands of spirals and nebulae. He discovered patterns among these objects, suggesting, according to his book *The Realm of the Nebulae*, a "classification [which] reveals a common fundamental pattern, whose continuous variation produces observed sequences of nebular forms." What Hubble produced in *The Realm of the Nebulae* was his classification and sequence that linked nebulae to spirals, including drawing comparisons of the Milky Way to Andromeda and M51.

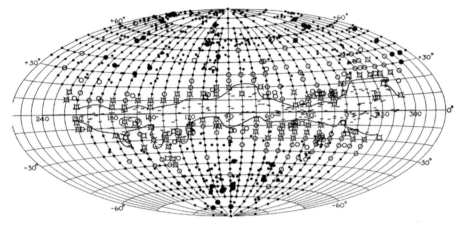

FIGURE 7.14 Distribution of extra-galactic nebulae (Hubble 1934).

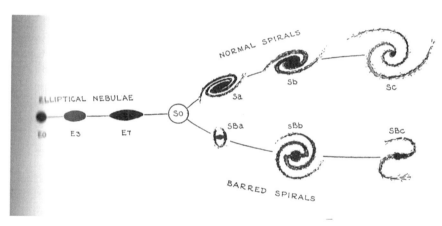

FIGURE 7.15 Hubble's "sequence of nebular types" (*The Realm of the Nebulae*).

According to Hubble, the Andromeda Galaxy was an SB spiral, and the Milky Way was either an SB or an SC. Hubble's classification appeared to solve several cosmological problems. First it provided a method for connecting nebulae to spirals, thereby refuting Shapley's argument that the two were distinct. This then provided yet another argument against the single universe theory. Second it also seemed to describe an evolutionary sequence for the universe, showing how different spirals evolved from nebulae.

Although today Hubble's evolutionary pattern has been rejected, in the 1930s it was consistent with ideas arising from Kant regarding how the creation of solar systems and island-universes may have evolved from accretion, rotation, and the eventual collapse of nebulae into spirals of some form. Thus, the Dumbbell Nebula was simply an early form of a galaxy, with the Sombrero and the Large Magellanic Cloud later versions, and the Andromeda and Milky Way galaxies still later versions. Hence, with Hubble's classification, all nebulae were potentially proto-galaxies, thereby again reinforcing the idea that all of these objects represented distinct non-Milky Way objects in the sky. This classification therefore also reinforced the island-universe theory.

In describing this evolutionary classification, Hubble could draw more comparisons between the Milky Way, Andromeda, and other galaxies, showing similarities that reinforced the island-universe theory. If the Milky Way were a galaxy in a pattern resembling other spirals, then the size and composition of other galaxies might not be much different than it is. What this meant then is that the structure of the Milky Way and, for that matter, the Andromeda Galaxy, might be models or patterns for other galaxies in the sky. By studying either of them one could gain insights into others. The stronger the parallels between the two, the better the two were as examples that could reveal information about the other and potentially all other galaxies.

Hubble's Legacy

Edwin Hubble's research and conclusions effected a second Copernican Revolution in astronomy. If the first Copernican Revolution displaced Earth from the center of the universe, the second one displaced the Milky Way from the center. It did that by demonstrating that there were other galaxies, potentially millions if not billions, beyond the Milky Way. Hubble demonstrated not only that the island-universe theory was correct, however. His research also significantly expanded the size of the universe. The distance to M31 was nearly 1,000,000 light-years. This considerably enlarged the universe over what Ptolemy thought, and over even what astronomers in the late nineteenth and early twentieth century believed. Hubble's universe was vastly larger and more populated with galaxies than previously thought.

Hubble's research effectively ended the Great Debate. It also produced a classification of nebulae that purported to show an evolutionary process in the universe. Nebulae were formed, rotated, condensed, and formed more regular spirals. Thus, the universe was not static but organic, demonstrating a history and pattern to it. Hubble's universe displaced humans even further away from the center. If the Medieval Christian Church's political theology was premised upon a Great Chain of Being, the chain clearly seemed broken.

Hubble's research changed astronomy, with the Andromeda Galaxy being at the center of it. Hubble's research transformed M31 into the paradigm or model galaxy from which information about all others could perhaps be gleaned. It was now perhaps the most important object to study if one wanted to understand the galactic nature of the universe.

Yet Hubble's classification and original calculations to M31 left unresolved several problems. Hubble was off considerably in terms of Andromeda's distance, diameter, and mass. For example, Shapely had argued that the Milky Way was 300,000 light-years in diameter. Hubble reached a different diameter. Who was correct? Hubble was also far off with the true distance to M31, so it would be the task of future astronomers to resolve this issue. Third, astronomers such as Van Maanen had contended, based upon Hubble's careful analysis of spirals (including M33 and M31) that they had internal motions. If in fact they were rotating they could not be so large, otherwise they would have to be rotating at exceedingly high velocities. If Shapely and Van Maanen were correct, these island-universes had to be small in comparison to the Milky Way, rendering questionable that they were distinct galaxies on a scale similar to our galaxy.

In the 1920s and into the 1940s Hubble and others addressed these problems. Hubble himself responded to Van Maanen by retracing the latter's steps and measurements. He found no angular rotation, attributing any rotation to measurement errors. Other astronomers such as Stebbins and Whitford in 1934 used the brightnesses and colors of M31 to ascertain its diameter. They concluded that the "known diameter of the Andromeda Nebula has been more than doubled in the direction north and south from the nucleus, and that this same ratio of increase applies to the apparent minor axis or width of the nebula." Their estimate of Andromeda being 20,000 pc in diameter compared favorably to the 30,000 pc estimate for the Milky Way. Williams and Hiltner in 1941 reached a similar conclusion, finding M31 to be 24,400 pc (80,000 light-years) along its major axis.

One last issue seemed to remain – to prove the true nature of Andromeda. Thus far M31's central region had not been resolved into stars. Lacking this resolution, one could argue that the spirals really were nebulae composed of gas and not stellar and therefore not island-universes. Technology and improved photographic techniques eventually ended this issue when Walter Baade in 1944 announced that he had resolved the stars of M31, as well as its companions M32 and NGC 205. Thus, Baade provided the final proof to resolve the Great Debate.

Yet Hubble failed to explain the redshifts and blueshifts of galaxies such as Andromeda. Were the galaxies rotating around their core or was something else taking place? This is the problem that Hubble and others now had to address.

8. Andromeda, Galactic Redshift, and the Big Bang Theory

Edwin Hubble's establishment of the Andromeda Nebula as a distinct galaxy demolished one vision of the universe that had existed in the West since the ancient times. His use of Cepheid variables in M31 to determine the distance to this Messier object not only provided definitive proof and support for the island-universe theory but it also forced astronomers to enlarge their estimates of the size of the universe by many times. If the Andromeda Galaxy, according to Hubble, was nearly 1,000,000 light-years distant, the universe was infinitely larger than any had previously thought. But was it truly infinite, or were there some bounded limits to its size? Moreover, what was the shape of the universe, and could one speak of it having a center? These were but a few of the questions that remained as the 1920s closed and the last remnants of the single-galaxy universe theory were abandoned and replaced by one defined by an island-universe view.

Hubble's work on Andromeda solved many questions, but not all. For example, how to explain the movement or velocities of the galaxies? Slipher and others had detected movement associated with galaxies. On one level this movement, at least for spiral galaxies, suggested that they were rotating around their core. This spinning indicated a process of galactic formation – the condensation and spinning of nebulae that eventually yielded irregular galaxies and then eventually spirals such as the Milky Way or Andromeda. But there also seemed to be another type of movement associated with these galaxies, and this was tied to their red- and blueshifts. Most galaxies demonstrated a redshift – an apparent movement away from Earth, while Andromeda itself evidenced a blueshift, indicative of it moving toward the Milky Way and our galaxy.

What did this mean? Were the galaxies actually in motion, moving away or toward Earth, or was the motion simply something associated with their rotation? This is a question that Hubble continued to address in the 1930s, when he concluded that the galactic movement was indicative of something more profound regarding the universe. What he concluded was that the universe itself was in fact expanding, and what he eventually discovered was perhaps one of the most famous numbers in all of astronomy – the rate of expansion of the universe, now referred to as the Hubble constant. His discovery that the universe was expanding solved many problems, including Olber's paradox, and perhaps the ultimate question: How was the universe created?

The Paradoxes of a Closed, Static Universe

Perhaps until as late as the twentieth century the prevailing belief in Western thought and astronomy, and perhaps also across cultures around the world, was that the universe was closed, finite, and static. This meant it was assumed to be of finite dimensions, with a fixed center, and that once created, it has remained the same until now. For the Judeo-Christian world, the origins of the universe are described in Genesis from the Bible. God created the heavens and Earth in 7 days. God is the ultimate creator of the universe. It was a universe that Bishop Ussher in the seventeenth century declared to have been created on Sunday, October 23, 4004 B.C. He also pronounced that Adam and Eve were expelled from the Garden of Eden on Monday, November 10, 4004 B.C. He reached his conclusion for both of these events, as well as others, by reading the Bible and calculating dates based on events reported in the scriptures.

The universe that was created, most believed, was static. Once God had finished with Creation, the universe was done. Earth, the Moon, the Sun, the planets, and the stars were all fixed in their places, never to change until the end of time. As discussed in earlier chapters, the Ptolemaic depiction of the cosmos reflected this fixed nature, with Earth at the center followed by increased concentric rings that housed the various objects in the sky. Earth stood still at the center of this fixed universe, except for the rotation of the cosmos around Earth. Fixed meant that no new objects were being created, none were disappearing. All cultures similarly held tightly to the idea that Creation was permanent or fixed. From ancient Egypt to China to the Indians of the Americas, their religions and folklore assumed a fixed and permanent universe. As a rule, visual observation seemed to confirm this assumption. But there were events that tested this belief.

For example, appearances of random meteorites, or worse, comets, were greeted with concern. At a time when astronomy and astrology were joined, the appearance of these objects was viewed as portending some major event. Even the appearance of the "Christmas star" foretold to the Three Wise Men that something special was about to occur. Changes in the sky were heavily scrutinized and recorded. Finding something out of the ordinary demanded explanation, since the appearance of a comet or another phenomena challenged the otherwise constancy of the sky.

FIGURE 8.1 M1, the Crab Nebula (Hubble Telescope, NASA).

For example, the appearances of supernovae were noted across cultures. Dating back to as early as the second century, Chinese astronomers recorded the appearance of supernovae, or so-called guest stars. There are indications that the Chinese also observed a supernova in A.D. 393. In 1006 a supernova in the constellation Lupus was recorded in China, Europe, Egypt, and perhaps in North America. But one of the most famous supernovae was the appearance of one in 1054, which was widely noted in China, Japan, and the Arab world, and perhaps among some of the North American Indians. It appeared in Taurus and was reputed to be four times as bright as Venus, lasting for about 3 weeks. The remnant of that supernova is M1, the Crab Nebula.

Two other famous supernovae visible to the naked eye were SN 1572, observed by Tycho Brahe in that year, and SN 1604, seen by Johannes Kepler in that year.

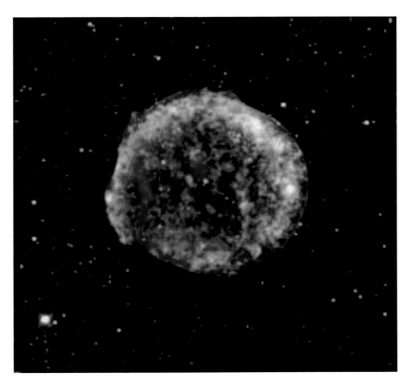

FIGURE 8.2 SN 1572.

Andromeda, Galactic Redshift, and the Big Bang Theory   185

All three of these supernovae were objects of curiosity for many reasons, but the principle one in the West and perhaps across cultures is that their appearance created a rupture in the permanence of the heavens. For a sky that was supposed to be permanent and unchanging, new stars were startling. Their appearance forced questions and doubt about the cosmos and a belief system built on it. Similarly, the invention of the telescope, Galileo's discovery of new stars, subsequent findings of new planets, and supernovae in Andromeda raised problems. The result of this discussion? It forced almost a return to the beginnings (especially in the West) to ancient philosophers and scientists, who tried to explain the problem of permanence and change in the universe.

Even with rejection of the ancient myths and depictions of the sky, including the Ptolemaic vision, several assumptions persisted. Isaac Newton, whose three laws of motion and theory of gravity changed much about the way astronomers thought about the universe, produced a mechanistic order to the cosmos, with change occurring as a result of a variety of forces interacting. But in a Newtonian view of the universe, space is infinitely static and flat. It is like a large piece of graph paper upon which stellar objects are hung. (Perhaps it would be better to describe it as a large cube with objects hanging in it).

These objects have no impact or imprint on space outside of occupying some. Similarly, time for Newton, as it had been for Christian Europe at least since the days of the fourth century theologian St. Augustine of Hippo, was one-dimensional. By that, we mean time began with Creation, moves in one direction, and does not change its pace. Finally for Newton, time and space do not interact. Time is experienced the same regardless of whether one is on Earth, the Moon, or on a distant star.

There are problems, however, with the Newtonian concept of space and time. One problem that a closed finite depiction of the universe posed was Olber's Paradox, which was discussed in an earlier chapter. Simply put, if the all the stars in the sky were emitting light, why was the night sky not as bright as the day sky? Assume that the sky is full of shining stars equidistant from the Earth. Their light should all reach Earth at the same time, especially if all the stars were created at the same time. Now perhaps at a time when the cosmos was not seen as containing too many

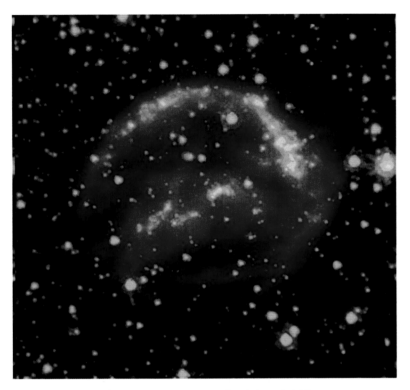

FIGURE 8.3 SN 1604 (Spitzer image in infrared, NASA).

stars one could address this question by declaring that there were simply not enough stars in the heavens to light up the sky. But once Galileo in 1610 discovered the Milky Way was composed of many more stars and subsequent astronomers saw the same across many parts of the sky, the paradox became even more difficult to explain. Recent Hubble Telescope pictures confirm that the universe is rich in stars.

Theoretically, point a telescope in any direction in the sky and one will find stars. If there are stars everywhere, all emitting light, they should fully brighten the night sky. But they do not.

Why? Hubble could only partly answer that question with his 1920s research on the Andromeda Galaxy. He could argue that not all galaxies and the stars within them are the same distance from Earth. At varying distances their light had to travel further than some to reach Earth. Thus the night sky is not as bright as the day sky because not all the light from the distant stars has reached

FIGURE 8.4 Hubble Telescope deep sky southern sky image.

Earth. But in time that would change and they would. Moreover, one could also resolve some of Olber's Paradox if one recognizes that some stars are older than others, with new stars being created and older ones dying off. Also, the existence of stellar dust obscuring stars and galaxies could also account for the darkness. Yet this model would then necessitate that the static model of the universe be modified; not everything had been created at the same time.

Einstein's Universe

Albert Einstein's impact on astronomy and how the universe is thought about cannot be underestimated. His equation $E=mc^2$, reflecting relationships between mass, energy, and the speed of light, may perhaps be the most famous mathematical formula in history. His is a name almost everyone recognizes, equating it with the concept of genius. The significance of Einstein' work was to vastly redefine the Newtonian universe. Along with that of Edwin Hubble and others in the early part of the twentieth century, his work completed the demolition of the older, flatter, and more static vision of the universe.

Let's discuss again Einstein's special theory of relativity, which was introduced in Chap. 5. What that theory states is that our experience of reality (space and time) is the same for those of us moving at a constant velocity. But there is a different experience of space, for example, if you observe an object move by you quickly (and you are motionless), which results in a length contraction. Conversely, fly at a high speed with a clock and compare how time ticks off compared to one located on the ground. The flying clock will tick off slightly less time.

The significance of Einstein's special theory of relativity was to alter some basic premises of the Newtonian cosmos. The first was to demonstrate that time and space are not distinct; they are connected in what is called a fourth dimension of space-time. Recognizing this new fourth dimension means that individuals and objects in the universe may not share the same experiences of time and space, as the Newtonian theory would suggest, if they are located in different spaces and moving at different velocities. This aspect of the special theory of relativity demolished the idea or premise that there was a center or focal or vantage point from which to observe or view the universe. For geocentrics, Earth was that center, but Copernicus destroyed that. For heliocentrics, the Sun provided that center. Einstein's special theory of relativity implied no center. Thus, the analogy of thinking about the universe as a cube or flat piece of paper falls apart. In both cases one can locate a central point, but for Einstein, no such place exists.

The special theory of relativity was only one way that Einstein challenged depictions of the cosmos. In 1915 Einstein also developed his general theory of relativity. This theory builds upon and corrects Newton's concept of gravity. Think about the famous (but possibly mythic) story of how Newton discovers gravity – an apple falls on his head. Gravity makes it appear that the apple is falling to Earth. But what if one replicated this phenomenon with the apple remaining stationary and Earth moved toward the apple? If that were to occur it would be difficult to conclude which object – Earth or the apple – were stationary. Einstein referred to this problem as the equivalence principle. This principle states that in a finite volume of space the downward pull of gravity can be replicated by the opposite acceleration of the observer.

The equivalence principle places the emphasis of the experience of gravity more on the motion than on force, and it also situates it within space-time. Einstein contended that to think of gravity within space-time meant that the former acted to curve space or that it pulled in such a way that time and space were dilated or distorted by it. Gravity is motion. Think of space as a gigantic flat grid. Now drop a heavy object into that grid. What it does is to distort the grid. That distortion affects space-time. It pulls space-time, giving it some curve or shape to it beyond being flat.

There are several ways to explain the significance of this observation about gravity within the general theory of relativity. Image the heavy object in the grid is a star. Light that approaches it is bent toward it.

Similarly, space is also bent or distorted. Now imagine a universe littered with many massive objects all distorting space-time. The result is a universe with lots of distortions and curves or bends in it. This distortion of space-time can thus account for several factors or observations. For one, gravity can bend light. Newton's theory did not predict this, but Einstein's can. A star observed not far from the Sun will be seen in a location slightly different from its actual position. Second, in the nineteenth and early twentieth centuries astronomers sought to create an accurate model of Mercury's orbit around the Sun. Newton's theory of gravity suggested that once the pull of the other planets on Mercury is taken into account an accurate prediction should have been possible, but it was not. This led many in the nineteenth century to hypothesize that another planet x – Vulcan – must be distorting Mercury's orbit. Einstein solved the calculation problem by invoking his general theory of relativity to demonstrate the planet's orbital path was distorted in space-time. Finally, the general theory of relativity suggests that gravity can create a redshift, distorting light waves and time similar to what occurs with a Doppler effect.

Einstein's theory of general relativity can also explain or predict a couple of additional effects. One is the idea that perhaps an object can be so massive that it would alter space-time around it.

This is what a black hole would do. A black hole is the product of a star collapsed unto itself. Gravity literally creates a sinkhole out of the star. This sinkhole then gobbles up objects close to it, getting more massive over time. But the space-time located near the black hole is curved, making it impossible for light to escape from the hole. Thus the name – black hole.

It has also been suggested that near the black hole time and space would be dilated. Studies have suggested this to be accurate. Finally, Einstein also contended that massive objects bending light would serve as large lenses able to magnify more distant objects. Again, studies have proven this to be accurate.

Overall, the special and general theories of relativity suggest a universe different from Newton's. It is curved, centerless, and distorted. Massive objects can produce redshifts and make objects appear where they may not be. The importance of these Einsteinian conclusions becomes clear when Hubble's work is reexamined.

Andromeda and Galactic Redshift

Recall the concept of the Doppler effect. This is the phenomena of sound appearing to get louder as an object approaches, decreasing as the object moves away. The Doppler effect occurs because as an object approaches, the distance between the wavelengths of the sound emitted become shorter for each crest. Because of the increased frequency of the sound wave, the object sounds louder as it approaches, quieter as it departs.

The Doppler effect is not confined to sound, as Christian Doppler and others in the nineteenth century discovered. It also applies to light. With the discovery of spectra and spectroscopy as a tool of analysis, light from distant astronomical objects could be examined, determining not only the chemical content of stars but also their movement in the sky. Redshifts indicated that objects were moving away from Earth, blueshifts toward Earth. The Doppler effect made it possible to detect movement and motion in the universe.

Vesto Slipher's groundbreaking work in the early part of the twentieth century on spirals revealed varying radial velocities for these objects. In 1913 he reported a blueshift for M31 as approximately 300 km/s. In 1914 he presented a paper arguing that all spirals appear to be rotating. He reached this conclusion again by studying the red- or blueshift of them. Subsequent papers described how far more objects displayed red as opposed to blue. Exactly what was the significance of these redshifts (or blueshift for M31) was not immediately understood.

In 1924 Swedish astronomer Knut Lundmark used observations and records of the shifts associated with the Andromeda Nebula in an effort to compute its distance from Earth. In the same article he also used this data to determine if it revealed a curvature of space-time in the universe. Edwin Hubble and a colleague, Milton Humason, built on this work and in a 1931 paper sought another means of ascertaining the distance to spirals beyond the use of Cepheids. Here they turned to the absolute magnitudes of the nebulae, keying in first on Andromeda and then other spirals. In the process of researching them they also noted a shifting of the spectral lines of these objects. In effect, there was evidence that these galaxies were moving, generally away from Earth. In this article they state that they would use the "'apparent velocity-displacements' without venturing on the interpretation and its cosmological significance." What were Hubble and Humason referencing?

Their extensive study of galaxies, starting with M31, revealed an interesting relationship. They found that the further the distance the galaxies were away, the greater the redshift, i.e., the more quickly they were moving. Expressed mathematically, they discovered the following relationship:

$$\text{Velocity} = \frac{\text{Distance (parsecs)}}{1,790}$$

At first in the 1931 article the relationship was described for 46 nebulae, while later on in 1936 Humason extended the study to include 100 additional velocities drawn from 6 clusters, 5 groups, and 56 isolated objects.

Humason's conclusion was that the velocity-distance relationship was "sensibly linear to a distance of 70 million parsecs."

FIGURE 8.5 Milton Humason.

As stated in a subsequent 1935 article by Hubble and Richard Tolman: "Light arriving from the extra-galactic nebulae exhibits a shift towards the red in the position of its spectral lines, which is approximately proportional to the distance to the emitting nebula." What did this relationship mean, and what was its cosmological significance? Some, such as Reynolds in 1938, studied M31 and other spirals and mused that it was a sign of galactic recession. By that, Reynolds contended that perhaps some galaxies were moving away from Earth, whereas others, including Andromeda, were moving toward Earth. Thus, by the middle of the 1930s, studies of red- and blueshifts of galaxies were beginning to raise questions about the velocity of the shifts, how they corresponded to distances, and whether they were signs that the galaxies themselves were actually in motion relative to Earth, and not just rotating. In effect, why were galaxies apparently moving toward or away from Earth? These were all perplexing questions.

The 1935 Hubble and Tolman piece offered two suggestions in an attempt to address these questions. One was that the redshift and accelerating velocities were due to some unknown cause such as stellar extinction (whatever that was) or absorption by space debris. These answers simply sought to explain away the data by contending that the recorded red- or blueshifts were false, produced as a result of some other phenomena interfering with data recording. Galaxies were not really moving, according to these explanations.

A second suggestion was that the shifts were the result of the spatial curvature of the universe. If space were curved the shifts were similar to optical illusions, such as when objects on Earth, such as boats, are observed traveling into the horizon and appear to fall off. Maybe some type of similar phenomena was occurring with a curved space. Distant spirals perhaps had their spectra stretched or shifted in a curved space. This latter was a conclusion similar to that reached in 1924 by Lundmark. Conceivably, Hubble and Tolman could have drawn on Einstein and the general theory of relativity to contend that the gravitational pull of massive objects was causing these shifts. Yet this would not have been an adequate answer. How could massive objects such as galaxies be red- or blueshifted by other objects such that it would affect their spectra? Maybe their own mass was the cause of the shifts? These were answers that did not make sense or seem to fit.

Yet the more powerful answer was one that assumed that the galaxies were actually receding and that these objects were moving away from Earth at velocities that increased with the distance from Earth. According to Mario Livio, in a 1931 piece in the Belgian science journal called the *Annales de la Société Scientifique de Bruxelles* (Annals of the Brussels Scientific Society) Georges Lemaître suggested perhaps that the universe was actually expanding, but his claims went unnoticed. It would be Hubble who would eventually get credit for the expanding universe thesis.

In making this argument, Hubble and Tolman did seek to reconcile their empirical work with that of Albert Einstein's claims regarding the universe and general relativity. However, Einstein's theory of special relativity seemed to be able to explain what was going on here: Emitted light from receding objects needed to be understood from the relative position of the Earthly observer in relation to the object, with those more distant appearing to move at greater velocities.

What Hubble and his various colleagues were detecting was in fact cosmological, more than something that could be explained by a theory of relativity; they were not observing simply a relative acceleration or velocity of galaxies as a product of the distance from Earth. Instead, this relationship between velocity and distance, eventually called Hubble's constant, revealed a universe that was actually expanding. This vision of the universe, for Hubble, was compatible with Einstein's claims. According to Hubble in his landmark 1936 book, *The Realm of the Nebulae*:

> Current theories of cosmology employ a model known as the homogenous, expanding universe of general relativity or, more briefly, as the expanding universe. It is derived from the cosmological equation which expresses a principle of general relativity – that the geometry of space is determined by the contents of space.

What was Hubble stating here? Quite simply, the universe was actually expanding, and its size and shape were a product of what was in it. The actual makeup of the universe – its stars, galaxies, and other objects – have an impact on the shape and size of it. Two examples can clarify this. One example is to assume the universe is a big box, a cube, with objects hanging from or located in it. The shape of the box (universe) and the objects (stars and galaxies, etc.) in it are not connected. The latter are simply situated in a predesigned box. As the box expands in size the objects stay where they are. This is a model Hubble rejects. Instead, from the very beginning the shape of the cube or box is determined by the objects inside, and as the cube expands so do the distances between the objects. In fact, the objects inside do not simply hang in the cube; they affect or influence the shape of the cube itself as it evolves. This is what Hubble was arguing. This model is consistent with Einstein's general theory of relativity – gravity can impact the shape of the universe itself.

A second analogy to explain what Hubble was arguing is to imagine the universe as an expanding balloon. Imagine dots on the balloon as representing galaxies. As the balloon expands the dots initially further apart appear to move apart faster than those that were initially closer to each other. But now somehow imagine that the dots on the balloon affect the shape of the balloon. As it expands its shape is affected by these dots. This, too, is what Hubble seemed to be describing.

From Hubble's original efforts to measure the distance to the Andromeda Nebula, he had discovered something more significant.

He had provided empirical evidence for island-universes, Einstein's theories, and now for an expanding universe. What he discovered was a relationship between the velocity at which a galactic object was moving and its distance from Earth. The further away a galaxy was from Earth the faster it was traveling. What emerged from this discovery was what has come to be known as Hubble's constant.

Hubble's Constant and Measuring the Universe

Hubble discovery of a galactic redshift was no less significant than his determination that Andromeda was a distinct galaxy. If the former discovery provided proof for the island-universe theory in a cosmos larger and more populated than previously thought, the galactic redshift demonstrated that we did not live in a static universe. Instead, it was ever expanding. Hubble's constant expresses that relationship. As it has subsequently been refined, the galactic redshift that Hubble formulated can be expressed as:

$$V = H_o d$$

where V equals the velocity of a galaxy, d equals the distance the galaxy is from Earth, and H_o is the Hubble constant. The formula, as noted above, finds a linear relationship – the farther away a galaxy is the greater its velocity. Comparing two galaxies, the one that is three times further away from Earth is receding at three times the velocity compared to the closer one. But what is the value of H_o, the Hubble constant? This has been the subject of significant debate and research in astronomy.

The Hubble constant expresses a relationship. For every x units of space away from the Earth a galaxy's velocity increases by a certain amount of kilometers per second. The Hubble constant is like the slope of a straight line on a graph. For those familiar with simple algebra, there is a simple equation often taught, $Y = mx + b$, as a measurement for the slope of a straight line. "m" expresses the slope. The Hubble constant is similar to mx. Calculating H_o is difficult because of the problems in securing exact measurements of distances to galaxies. Use of Cepheid variables to measure distances to galaxies is accurate but only up to a certain point from Earth. Beyond about 30 MPC (mega parsecs) from Earth Cepheid variables are too dim and fade. Other measurement tools to ascertain distances, such as the use of other variable stars, supernovae, or spectral analysis, also have limits in terms of their accuracy beyond a certain distance from Earth.

Hubble himself originally estimated the velocity/distance relationship to be 500 km/s/Mpc. That number has been significantly refined and downgraded since. Allan Sundage, Hubble's successor at Mount Wilson and then the Mount Palomar observatories, argued in a 1956 joint paper with colleagues that the Hubble constant was 180 km/s/Mpc. In 1958 Sundage alone argued it was 75 km/s/Mpc, while by the early 1970s he and others had reduced it even further to 55 km/s/Mpc. Conversely, astronomers Sidney Van den bergh and Gerard de Vaucouleurs in the 1970s concluded the Hubble constant to be approximately 100 km/s/Mpc. Controversy raged over the next several decades in determining the value of H_o because of difficulties in securing precise measurements to distant galaxies. However, with the launch of the Hubble Telescope in 1990, more precise measures of Cepheids have solidified valuation of the Hubble constant to be approximately 70–72 km/s/Mpc. Not all astronomers agree on the exact value of H_o, but this 70–72 km/s/Mpc seems to be the most widely accepted estimate today.

The Significance of the Hubble Constant

Hubble established that the universe was expanding, but he was not the first to do that. Einstein, too, had at one time proposed a constant in order to reconcile his observations with his general theory of relativity. Initially Einstein bought into the Newtonian idea that the universe was static, but his calculations did not work. He thus proposed what he called a cosmological constant. This constant would be a force equal to that of the gravitational pulls that would force the universe to collapse upon itself. This cosmological constant was almost like a "fudge factor" for Einstein, inserted to make sense and reconcile his arguments with a Newtonian static universe theory. Later in life Einstein said of the cosmological constant that it was the "greatest blunder of his life." However, had Einstein rejected the Newtonian static universe theory and opted for an expanding one the constant would not have been so blundering. Einstein was on the right path; he just did not see the implications of the cosmological constant in terms of its significance.

The Hubble constant, whatever its exact value, leads to several important questions and implications. First, does the value establish or contradict Einstein's specific theory of relativity? The fact that galaxies are receding from Earth seems to suggest they are moving away from our planet, thereby establishing some notion that the Milky Way (or at least our Solar System within it) is at the center of the universe. Two responses are possible to reconcile with the specific theory of relativity. One, the recessional appearance would look the same from any other point in the universe. Galaxies are receding from one another at a constant linear velocity and thus the same observations and calculation for the Hubble constant would be ascertained from anywhere in the universe. It only looks like we are at the center of the universe because this is the point from which we are observing. Go to another galaxy and observe the Milky Way, and it would appear to be moving away from it. Second, not all galaxies are receding from the Milky Way. As noted, the Andromeda Galaxy is actually moving toward Earth, pulled toward it in part by the gravitational pull of the two upon one another. In the distant future, Andromeda and the Milky Way will collide and merge. Thus, galactic redshift can be reconciled with an Einsteinian view of the cosmos.

A second issue is how the Hubble constant and relationship establish the concept of a cosmological redshift. The cosmological redshift is caused by the expansion of the universe. This is not the movement of objects in space. It is the actual expansion of space or the size of the universe. Astronomers have constructed a formula to determine how the universe has expanded over time.

$$Z = \frac{\lambda - \lambda_o}{\lambda_o}$$

In this equation, z equals the redshift, λ_o equals the unshifted wavelength that is observed, and λ is the observed wavelength. The λ/λ_o ratio represents the amount of wavelength stretching or elongation observed. This equation is important because if one can determine Z then one can calculate its velocity. Once velocity is known, one can then begin to estimate distance for a galaxy. Thus this measurement of redshift corresponds to both velocity and distance, thereby giving astronomers a better tool for ascertaining how far away objects are. But why only an estimate?

Hubble's constant is not really so constant. In addition to astronomers not being able to pinpoint its exact velocity, observations of very distant galaxies indicate that the universe's expansion rate has accelerated over time. How do scientists know this? In observing the brightness or luminosity of supernovae, astronomers have found that the acceleration curve for the universe seems to be increasing over time. For about 2 billion light-years Hubble's constant has been linear, but further than that it appears slower. Drawing a graph over very large distances of billions of light-years the linear line seems to accelerate. Why is this?

It is important to understand again the relationship between light, distance, and time to explain this. To say that an object in the universe is 2.5 million light-years distant, as the Andromeda Galaxy is approximately from Earth, means that it would take light traveling at 3×10^5 km/s, 2.5 million years to reach our planet. Stellar or cosmological distance is equated with time. This is another way of recognizing that space (or distance) and time are connected. If one were to go out tonight, observe the night sky, and see M31, the light or image that one is seeing left the stars of that galaxy 2.5 million years ago. For the Sun, the light rays that reach us left about 8 min ago; for Proxima Centauri, the closest star to us (besides the Sun), the image or light seen is 4.2 years old. Were this star to explode and become a supernova today, we would not know of that fact for 4.2 years. Similarly, were anything to happen today in M31, it would be another 2.5 million years before we would know that.

For all objects, but especially for very distant ones, what astronomers see in the sky tonight is something that occurred millions if not billions of years ago. To look further and further distant into space means one is also looking back in time. What

FIGURE 8.6 Proxima Centauri.

appears to be the case is that the more distant in the past we go, to a time when the universe was much younger than it is today, the slower the cosmological redshift. The universe initially expanded, and over time the value of the Hubble constant has increased.

However, all of this raises several new questions. Why this acceleration? How big and, related to that, how old, is the universe? And perhaps the ultimate question – what does all this suggest about the origins of the universe?

The Origins of the Universe

In a branch of philosophy called metaphysics (related to ontology) the most basic question to ask is why there is something rather than nothing. Why does anything exist, perhaps including the universe? Ancient cultures, as the opening chapters of this book, discussed some stories about the structure of the cosmos that also provided explanations for the creation of the universe. Every culture and religion seems to offer a story about creation. For the

Judeo-Christian tradition it is the story of Creation as outlined in Genesis, where God creates the universe, including Earth and humans in 6 days (the seventh day is rest).

The Genesis story and Christianity support a story where the universe comes into creation that eventually ends at some future millennial point. God thus provides the explanation for why and how the universe was created with some, as discussed before, calculating the age to be dated from 4004 before Christ. Given the finite nature of the universe, Martin Luther (1483–1546), a German theologian who broke away from the Roman Catholic Church and helped fuel the Protestant Reformation, was reputedly once asked what God did before the universe was created. Supposedly he replied that God was constructing punishments for those who ask questions like this!

Regardless of what Martin Luther thought or supposedly said, he, Christianity, and almost every culture and religion assert that the universe did have a beginning. But when and why? How did universe begin, and how does that origin speak to the acceleration of Hubble's constant?

If the universe is accelerating and increasing in size over time, then if one moves backwards in time the universe ought to have been smaller. The cosmological redshift implies that if the size of the universe has increased with the passage of time as it progresses, then at an earlier point in time it was smaller. Trace this back further and further in time and one reaches some initial point when the universe was exceedingly small and compact. At this point in the past the entire mass of the universe was pressed into one small and infinitely dense point. From this the universe then expanded. Something happened to move the universe from this infinitely dense and compact point to begin the process of cosmological expansion. Astronomers call this initial starting point and theory of the creation or origin of the universe the Big Bang theory.

The term Big Bang was initially given to this theory in 1949 by British astronomer Fred Hoyle (1915–2001). The actual origins of a hypothesis about the universe starting from some initial point can be traced to Georges Lemaître (1894–1966), a Belgian priest and astronomer who taught at the Catholic University of Louvain in Belgium. In trying to ascertain a theory about cosmic evolution that took into account the work of Hubble and the cosmological redshift or expansion of the universe, he contended in 1931 that it began with what he called a primeval atom. The genesis of this idea came from research and observations on radioactive decay. He had noted that some atoms, such as uranium, break down over time into smaller ones. New smaller atoms are formed and, in the process, radiation and energy are produced.

Lemaître hypothesized that a similar process might have occurred with the universe. From some primeval atom smaller components of the universe were produced, as well as energy and radiation. Evidence for Lemaître's analogy to radioactive decay also came from research by an Austrian physicist Viktor Hess (1883–1964), who in 1912 flew a balloon to a height of 6 km and recorded high energy particles, presumably coming from outer space. This was a startling result because physicists had contended that radiation would decrease as one traveled further from Earth. The reason was that it was assumed that all radiation was sourced in the planet. Hess's data suggested that the radiation must be coming from somewhere else besides the Earth – perhaps from outer space. Robert A. Millikan (1968–1953), an American physicist, in 1925 confirmed Hess's work and coined the term "cosmic rays" as the name for this radiation. If cosmic rays did exist, according to Lemaître, that was evidence of his primeval atom because this radiation was a by-product of the original Big Bang.

Finally, Lemaître also built on the work of Einstein. First, Einstein's famous $E=mc^2$ proposed a relationship between mass and energy. If the cosmic rays were one product of the Big Bang, mass would be another. Mass did exist in the universe in terms of things such as stars and planets. And thanks to Hubble's research on Andromeda, confirmation of the island-universe theory also proved the existence of galaxies. They were the eventual mass of the universe, built up from the initial primeval atom.

Einstein, in proposing his general theory of relativity, assumed that the universe was both homogenous and isotropic. By that we mean that the universe looks the same no matter where one observes, and the laws of physics apply the same everywhere. Lemaître used the Big Bang theory to account for these principles. A homogeneous and isotropic universe could only be possible if everything began from some initial starting point, exploded, and then spewed or emitted mass and energy in a uniform and isotropic fashion outward, expanding space and time as this occurred. Why suddenly the Big Bang occurred, overcoming gravity to hold everything together, is not understood. In fact, astronomers contend that the laws of physics as understood today did not exist. Max Planck (1858–1947) was a German physicist famous for the origins or quantum mechanics, a theory of subatomic particles. He and others suggested that it was not until a very short time after the Big Bang that the laws of physics that we observe came into existence. This time, 10^{-43} s after the Big Bang, is referred to as Planck time. From this instant on normal rules of physics exist.

Evidence of a Big Bang

Lemaître proposed an interesting theory, but what evidence supports it? If he is correct, the initial early universe should have been much hotter and denser than now, and it also should display evidence of some type of radiation or energy having been emitted. All three of these phenomena have now been verified.

Think of the primeval atom as a point of singularity from which the universe was formed. Everything – space, time, matter, and energy – are all together in this point. Once the Big Bang occurred the universe became incredibly hot, hotter, in fact, than at the core of our Sun, such that the subatomic particles fused into elements such as hydrogen and then helium. The universe was filled effectively with nuclear reactions, at least according to a theory proposed by Ralph Alpher and Robert Hermann around 1949. A similar argument was made in 1960 by physicists Robert Dicke and P. J. E. Peebles, who contended that the abundance of helium in the universe could be explained by these nuclear reactions.

Another offshoot or byproduct of this early universe and high temperature should be the existence of a cosmic microwave background – background radiation left over from the early years of the universe. If this radiation could be confirmed, that would provide evidence for the Big Bang. Such evidence has in fact been discovered, most conclusively with the launching of the Cosmic Background Explorer (COBE) by NASA in 1989, which found samples of this radiation.

Evidence of decreased heat or temperature has also been established. If the Big Bang theory is correct, and it was initially much hotter in the universe than it is now, then one should see a declining temperature curve over time. It would also suggest a time in the past when the universe was so hot that atoms could not form. Some cooling off period existed when the universe was so hot it was opaque. If this theory about the era of recombination is correct, the early years of the universe cannot be seen because matter is obscured. After an initial cooling off period the universe would allow for some combination or recombination of atoms. Thus, prior to this cooling off period the universe should have displayed high levels of energy and therefore high levels of temperature. The point when the temperature dipped to be cool enough for recombination was 3,000 K. This was approximately 300,000 years after the Big Bang. Evidence for this drop in temperature, or decrease in wavelengths for energy, was obtained by NASA in upper atmospheric balloon experiments in the 1990s. These experiments were able to detect variations in the microwave background radiation.

Third, if the Big Bang theory is correct, the average density of the universe should also display a downward curve over time. If all matter and energy were initially concentrated in a primeval atom, then after an explosion and cosmological expansion, matter should be dispersed further and further away from other matter, thereby yielding a less dense universe over time. At the point of the primeval atom, matter should have been so compacted that the photons that did exist were also packed together very tightly. All of this density meant a point where energy exceeded the amount of matter that existed. But after the Big Bang some point would emerge where temperature and energy would decrease, mass would increase (as consistent with Einstein's $E=mc^2$), and from there one would be able to detect a continuous decrease in density in the universe. Again, experiments in looking at temperature changes in the early universe, as well as evidence from the cosmic microwave background, confirm this too.

Overall, the evidence for the big bang theory, while not conclusive, is significant. Astronomers still do not know why the initial big bang or what happened before the emergence of Planck's time, but the big bang theory as originally proposed by Lemaître seems to hold up.

So if the big bang theory is correct, how old and how big is the universe? Growing out of Hubble's work one could then derive estimates of the age of the universe by correlating red shifts with recessional velocities to the most distant objects. If one calculated the red shifts far enough back to the most distant objects, back to a point a point of singularity, astronomers have estimated that the universe if approximately 13.7 billion years old.

Finally, why is the universe accelerating over time? Why is it expansion increasing as time progresses? One possible argument is that as average density decreases (as matter gets further and further apart) the gravitational attraction that holds them together decreases, thereby allowing for an acceleration of the universe's expansion. Other explanations look to the composition of the different types of energy and matter in the universe, distinguishing between visible matter and dark matter and the same for visible and dark energy. (Discussion of these topics occurs in Chap. 9.) However, the exact causal explanation for this acceleration is still in some dispute, and the big bang theory, at least as originally proposed by Lemaître has undergone modification as some questions with it have emerged and evidence or data gathered have posed problems.

Solving Olber's Paradox

So why is the night sky not as bright as the day? If the sky is infinitely rich in stars in every direction, the light from them should fill up the sky. The big bang theory and the concept of an accelerating universe help answer the questions. The most important answer is that not enough time has passed for the light from distant stars to reach us. When the universe was 3 billion years old its radius was similarly 3 billion light-years. A most distant star would have then also been nearly 3 billion light-years away. But now with cosmic expansion such a star may be nearly 14 billion light-years away. Its light thus has not had enough time to reach the Earth, and with an accelerating universe it may be possible that it never will.

The second reason is the cosmological redshift which Hubble discovered. It means that as distant light is redshifted, the wavelength increases and the energy decreases. The decreased energy due to the redshift means less brilliance or brightness and therefore less ability to brighten the night sky.

Finally, the big bang theory proposes theories about background energy and that may be obscuring the sky, thereby also making it difficult visible light to reach the Earth. This is similar to the phenomena of stellar gas and dust at the core of the Milky Way obscuring part of the night sky (and which also contributes to some blocking of the light from distant stars).

The Andromeda Galaxy in a Hubble-Einstein Universe

By the time shortly after World War II the old universe of Ptolemy and Newton had completely disappeared. A centerless, accelerating, expanding universe created from a point of singularity had replaced the older static one. What made this change in the depiction of the universe possible were Hubble's original observations and measurements of the Andromeda Galaxy along with Einstein's theories about relativity. This new universe was dynamic, accelerating, and truly infinite. It represented a new universe far different than the ones of the Ancients, the medieval Christians, and that of the astronomers of the beginning of the twentieth century. M31 was part of this new universe, as a galaxy born of a big bang.

9. Andromeda, Cosmology, and Post-World War II Astronomy

The Andromeda Galaxy and Hubble's Errors

Edwin Hubble's contributions to astronomy were immense. He resolved the island-universe debate, established a classification system for galaxies, and discovered that the universe was expanding. His work paved the way for the Big Bang theory. But Hubble was not the last word in galactic research. Several issues needed correction or completion, and much of this was addressed during and after World War II.

Astronomy after World War II benefitted from many new technologies that made it possible to view the cosmos in new ways. This included bigger telescopes, instruments that allowed for observation in other than visible light, and the placement of these instruments on orbital observatories above Earth. (Chap. 10 will discuss these new technologies.) However, the main task that dominated Post World War II astronomy was significantly filling in the details left over from Hubble, Einstein, and Lemaître. Astronomy sought to provide evidence for the Big Bang, perfect galactic measurements, and sketch out the implications of the general and special theories of relativity when it came to understanding the universe. But in doing so, astronomers also made many new discoveries. The Andromeda Galaxy was central to much of this research and work through the end of the twentieth century.

Hubble's Errors

One task of left incomplete by Hubble and others before him was resolution of individual stars in the Andromeda Galaxy. During the nineteenth century the failure to resolve individual stars in M31 left open questions regarding whether this spiral was simply gas and interstellar dust or a galaxy. Hubble, too, was unable to resolve the Andromeda Galaxy's stars. This task had to wait until 1944. That year Walter Baade (1896–1960) was the first to resolve the individual stars in the central region of the Andromeda Galaxy, perhaps providing the final if not belated proof of M31 as a distinct island-universe. Baade achieved this resolution with the same Mount Wilson telescope that Hubble had used. Why was he more successful than Hubble at this task? There were two reasons. First, he had at his disposal new filters and emulsions that enhanced photography. Second, Baade benefitted from the night time blackouts of Los Angeles during the Second World War, giving him darker skies than even Hubble had. As a result Baade was able to secure more light-gathering power and resolution with the Mount Wilson telescope than Hubble had been able to achieve.

Baade's resolution of the core of the Andromeda Galaxy, however, was not his sole contribution to either astronomy or knowledge

Figure 9.1 Walter Baade.

about M31. In seeking to resolve the stars in Andromeda's central region he succeeded in this task by using red-sensitive emulsions and filters. What he found in the core were red giant stars that were not detected with the blue-sensitive emulsions and filters used before. The latter had been employed because the stars along the periphery that had already been resolved were blue. Thus he discovered that at the core there generally appeared to be one kind of star, while in the periphery another. This was an important discovery leading to a distinction between what is now known as Population I and Population II stars.

Population I stars are relatively young stars. They are metal-rich. Astronomers use the word "metal" in a unique way, referring to stars that contain elements beyond hydrogen and helium. The Sun is a Population I star. Population II stars are older, metal-poor stars. Population II stars are larger stars, formed closer in time to the Big Bang. These stars are more massive than Population I stars, and having burned for so long, they have begun exhausting their cores, eventually increasing temperatures and producing heavier metals in the process. When these stars die and shed their gases and materials, their metals become the building blocks for Population I stars, as well as other objects such as planets.

Baade's analysis of the Andromeda Galaxy and conclusions about Population I and II stars is important in many ways. First, it added to the classification and understanding of stars in the H-R (Hertzsprung-Russell) diagram, at least in the sense of locating some stars temporarily in terms of their development. Second, the classification established further evidence for the Big Bang, providing an account of some stars being older than others. It did that by describing how older Population I stars were composed of the basic building blocks of H and He, and it is from the aging that heavier metals were eventually produced. Thus, this model accounts for the creation of original stars and then an evolution of new stars made out of the remnants of the Population I stars.

Moreover, this model also provides an explanation of how other matter in the universe was created. From the Population I stars Population II are created, along with the material for nebulae and rocky planets such as Earth. Finally, this classification indicated why certain types of stars were found in certain regions. Eventually the argument would be that older stars are generally

FIGURE 9.2 The Orion Nebula, M42.

FIGURE 9.3 Hubble image of the Eagle Nebula.

found in globular clusters, whereas new stars are located in the leftover gases and nebulae of older stars. The Orion Nebula, or M42, for example, is both a death and birth scene.

Orion is a nebula and the leftover material of older stars, but it is also the incubator of new stars. The Eagle Nebula, or M16, is similar in being a death and birth chamber for stars. Hence, Baade's exploration of Andromeda provided critical evidence to describe stellar and cosmic evolution.

After Baade, W. A. Baum and Martin Schwarzschild in 1955 examined both M31 and its companion NGC 305 in order to understand the source of their brightness and luminosity. In this study they investigated the brightness in relation to the star count. They found that while for NGC 305 the count-brightness ratio was consistent with that found in globular clusters, with M31 it was more consistent with that found in other nearby spirals. The significance of this was that much of the light in Andromeda was coming not from new Population II stars but old Population I ones. This research confirmed many of Baade' points, providing more evidence for how Andromeda has been an important source of information for understanding H I and H II regions, areas critical to explaining star formation and aging.

Baade's efforts to resolve the stars in Andromeda also led to two other corrections of Hubble's research and conclusions. Instrumental to Hubble's conclusion that M31 was a distinct galaxy was his claim that it was nearly 1 million light-years distant. To reach this conclusion he drew upon the research that Henrietta Leavitt did with Cepheid variables, using ones that he indentified in the Andromeda Nebula to calculate the distance to it. Baade examined the variables in M31, but he compared them to those found in the globular clusters located near the Milky Way. He found that in the clusters the brightest Cepheids had periods of 30–40 days, but similar period ones in Andromeda were much fainter than they should have been. This was especially true with the Population II stars he was observing. In an important 1952 paper, Baade resolved the controversy by distinguishing Type I from Type II Cepheids. The former were metal-poor, the latter more metal-rich. The period-luminosity is different for the two. Given the difference in periods, Hubble had erred in his calculation of the distance to the Andromeda Galaxy. Baade's estimates using the new Cepheids pushed the distance to M31 to more than 2 million light-years away, much closer to present-day estimates than were Hubble's. Baade thus doubled the distance to Andromeda and with that, at least doubled the size of the universe.

Hubble and Galactic Origins

Hubble had also been wrong on one other major point – the evolution of galaxies. In the *Realm of the Galaxies* Hubble had produced a tuning fork classification of galaxies. This classification began with irregular shaped galaxies and ellipticals, leading to a split in terms of spiral and barred spiral galaxies. Initially, this classification for Hubble also described an evolutionary process for galaxies. He assumed that as galaxies rotated and condensed they would evolve from irregular shapes eventually into a type of spiral. But there is a problem with this claim – elliptical galaxies do not rotate. It would be impossible for a galaxy suddenly to begin rotating or moving; that would violate one of Newton's laws of motion that bodies at rest tend to stay at rest. However it is easy to see how Hubble made this mistake.

There were two assumptions that led Hubble to this wrong conclusion. One dealt with galactic rotation, the other with arguments about the origins of galaxies. First, in the early part of the twentieth century many astronomers confused galactic or cosmological redshifts with galactic rotation. Spectrographic analysis of galaxies suggested they were rotating hundreds of kilometers per second. However, for many of these galaxies, the redshift observed was actually galactic movement – an enlargement of space-time and a receding of galaxies away from one another and the Milky Way.

Second, an important theory about the origins of stars originated in the writings and arguments of Kant and Laplace. They had argued that the origins of solar systems could be found in nebulae. As they rotate they condense. This rotation and condensation eventually produces a bulge at the center and then a flattening out along the edges. The picture sort of looks like a rotating pancake.

The central core of such a structure yields a star, with the additional material left over providing the matter for the planets and other solar system objects. This theory is one of two rival ones explaining the formation of solar systems, stars, and planets.

The other, a model based on accretion, contends that at least the planets in the Solar System were formed by accretion of smaller chucks of matter that bind together via gravity as they rotate around the central core. Today, evidence for both theories can be found, and they provide rival and in some case complementary theories about solar system and stellar origins.

However, the Kant-Laplace condensation theory also had some intellectual or theoretical pull when it came to galaxies. Perhaps galaxies formed as large masses of stellar gas or nebulae rotated and condensed. They then formed irregular-shaped, then elliptical, and eventually spiral galaxies as they continued to rotate and condense. This theory for galactic formation, along with evidence of their redshift, provided a basis for Hubble's theory of evolution.

Still, the theory is incorrect. As noted above, ellipticals do not rotate, or at least the rotation is slight. In addition, the Kant-Laplace theory might offer a convincing theory to explain the origins of stars, but it would be harder to apply this theory to galactic evolution. It would have to assume that there is a central mass at the core of all galaxies, and not simply a collection of stars. In reality, galaxies do not have a solid core. Third, the Kant-LaPlace theory would also have to account for the formation of billions of stars within each galaxy, with each of them then forming as a result of rotation and condensation. All of this would be impossible because it would then assume billions of smaller distinct rotations and condensations occurring within a larger galactic size rotation and condensation. These assumptions seem reminiscent of Ptolemaic cycles and epicycles to account for planetary movement!

Current theory treats Hubble's tuning fork diagram as a classification system and not an evolutionary theory. In that regard it is still a useful tool for classification and description, but not a good theory on the origins of galaxies.

Galactic Formation, Dark Matter, and Energy

If Hubble was not correct when it came to galactic formation, structure, and evolution, how did they get created? The Andromeda Galaxy factors prominently in explaining galactic origins.

FIGURE 9.4 Vera Rubin.

Let's look again at the idea of galactic rotation. This is what American astronomer Vera Rubin (1928–) did. Earlier in the twentieth century astronomers such as Slipher and Pease had contended that galaxies were rotating. In some cases there is a rotation with spirals, but they had also confused rotation with a galactic or cosmological redshift. There was thus an assumption that galaxies did rotate.

In the early 1960s Rubin became interested in M31 and galactic rotation. She was aware of the 1916 work by Pease, who had measured the galactic rotation of Andromeda. When he performed his analysis Pease needed 84 h over a 3-month period to obtain a spectral analysis of Andromeda in order to record its galactic rotation. By the 1960s technology and instruments had improved, and the time to record a spectrum had been reduced by nearly 90%. Rubin's goal was to compare M31's galactic rotation as far as possible away from the core. She anticipated that the rotational velocities would be different. Why? If one looks at the velocity of the planets rotating around the Sun, the further they are the slower their orbit. This is perfectly consistent with Newton's laws of gravity. A planet three times further away from the Sun would orbit the Sun at one ninth the velocity of the closer planet. There is a curve between distance and velocity. Thus, assuming Newton is correct; examination of the rotational speed of the Andromeda Galaxy should have demonstrated that regions further from the core orbit at velocities slower than velocities observed closer to the core.

However, this is not what Rubin discovered. Instead she, along with her co-researcher W. Kent Ford, found that there was no curve – it was a straight or flat line between distance and velocity. The most distant stars of the Andromeda Galaxy rotated at the same velocity as did the core. To a surprised crowd at a 1968 meeting of the American Astronomical Society (AAS), she announced her results.

In a 1970 article Rubin and Ford reported their results. In that article they also drew some comparisons between M31 and the Milky Way in terms of rotational velocities, finding parallels between the two. Yet their results for the Andromeda and Milky Way galaxies were not a complete surprise. Newton's laws notwithstanding, radio astronomy observations of Andromeda had detected similar results. Observations of other galaxies produced mixed results. None of this seemed consistent with the laws of gravity. For the outer regions of Andromeda to be rotating at the same velocities as the core, there had to be more mass in M31 than observed. But where was the missing mass?

The missing mass in Andromeda is related to another problem dating back to the 1930s that involved accounting for the galactic redshift and expansion of the universe. In turn this problem was related to another problem – explaining the clustering of galaxies. The single universe theory held that there was only one galaxy – the Milky Way – and Andromeda was part of it. Hubble's research refuted this claim, establishing M31 as a distinct galaxy. So far all this is correct. However, Andromeda and the Milky Way, although distinct galaxies, are not as separate as Hubble contended. Instead, from the 1930s on astronomers concluded that galaxies tend to cluster together. For example, there is the Local Group, composed of the Milky Way, the Large and Small Magellanic Clouds (which are really irregular-shaped galaxies), Andromeda, and several others totaling about 30. In addition to the Local Group, there is also the Virgo cluster, which includes two well-known spirals M84 and M86.

A more distant cluster is the Coma cluster, located about 900 million light-years away. Visually it is found in the constellation

FIGURE 9.5 M84.

FIGURE 9.6 M86.

FIGURE 9.7 Coma cluster of galaxies.

of Coma Berenices. Astronomers have also found that many clusters are connected into superclusters.

The clustering of the galaxies presents two problems: Why the clustering, and then how to explain the movement of the galaxies within these clusters.

Astronomers have suggested several possibilities to answer the first question. One has been to contend that when the Big Bang occurred the matter and energy that spewed out was rippled. By that, we mean it was not a smooth or even distribution of matter and energy. It was clumped. That clumping eventually meant that matter would form in chucks (hardly a scientific term!) and form galaxies that are connected to one another.

Whatever the reason for the clumping, galaxies seem to be connected in clusters. Evidence of that connection could be found in other established facts. One was that the galaxies seemed to exert a push and pull on one another. Andromeda and the Milky Way seem to have some type of gravitational connection, pulling the two together. The blueshift of M31 toward the Milky Way and the eventual collision and merger of the two in 6 billion years offers proof of that. Moreover, many other galaxies have already merged together, a product of gravitational attraction. In fact, astronomers have recently concluded that M31 is the product of a merger of at least two galaxies. Moreover, Andromeda has two satellite galaxies, M32 and NGC 205 (M 110), again further demonstrating galactic connections.

Another fact demonstrating a connection among galactic groups seems to be how they move together in space. They must be connected for such unison to occur. But this very connection is puzzling. If in fact they are connected and exert the pull on one another as observed, there needs to be sufficient mass within them for Newton's gravitational laws to explain this. Yet the mass is missing. This missing mass in galaxies was noted as early in the 1930s by Fritz Zwicky (1898–1974) and Jim Peebles in their book *Physical Cosmology*. Along with Jeremiah Ostriker, they did modeling of the Milky Way and other galaxies and found that the models of galactic rotation did not make sense given the mass that they could visually observe.

Zwicky, a Swiss-born astronomer, declared that the discrepancy between the observed and predicted mass was significant. He contended in 1933 that there had to be some type of *"dunkle*

FIGURE 9.8 Fritz Zwicky.

materie" (German for dark matter) that existed that accounted for the attraction among the galaxies in the Coma cluster. Zwicky was thus the first to declare that beyond visible matter, some dark or hidden matter existed in the universe, including within galaxies. He reached several other significant conclusions regarding galaxies. For example, he proposed, following up on arguments by Einstein, that galactic clusters could serve as gravitational lenses. This claim would eventually be substantiated in the 1930s. Zwicky also did landmark work calculating the distance to galaxies, and he produced with colleagues an impressive six-volume catalog of galaxies.

If dark matter did exist, including in M31, it could account for the gravitational hold in that galaxy and therefore explain the rotational velocities between the central region and periphery of that galaxy. Think of it this way: If M31 was not a collection of stars with lots of void or empty space in between, but instead dark matter connected to the stars, then it would make sense why stars at various distances from the core moved at approximately the same velocity. They would be connected like blades in a circular fan.

The existence of dark matter could also be used to explain three other problems related to the shape of the universe, the cosmological constant, and whether the Big Bang goes on forever. Although Einstein correctly argued that there is no center to the universe, this assertion did not address whether there was a shape to the universe. However his general theory of relativity noting the impact of gravity upon space-time indicates that some type of curving should occur. Gravity will curve space-time. But how much curvature there is in the universe depends on the amount of mass that exists in it. Depending on what astronomers call the density parameter of the universe, Ω_0, three results are possible.

One possibility is that the universe is flat and not curved. For example, imagine shooting parallel light beams into space. After traveling millions or billions of light-years they remain parallel. This is evidence of a flat and not a curved universe. However, if the beams converge this is proof of a positive curvature or a spherical universe. If they diverge, this is a negative curvature with a shape more like a horse saddle. Astronomers refer to the second and third possibilities as closed and open universes.

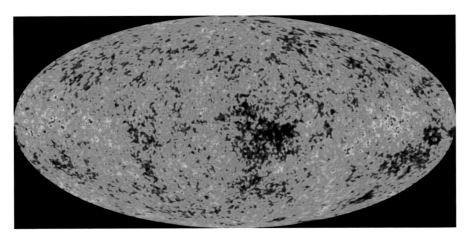

FIGURE 9.9 The Wilkinson microwave anisotropy probe depicting the cosmic microwave background.

The density parameter is connected to what is called the critical density of the universe, which relates Hubble's constant to the universal gravitational constant. All of this is also related to the average mass density of the universe. According to astronomers, three possibilities exist:

If $\Omega_0 > 1$, the universe is closed;
If $\Omega_0 = 1$, the universe is flat;
If $\Omega_0 < 1$, the universe is open.

Research, including that on the Cosmic Microwave Background to see if its energy or light is bent, has tended to support a flat universe theory. But the flat universe creates a problem.

There is not enough visible mass to support the flat theory. Something is missing. This gets to a second problem, the velocity of the cosmological constant. How does one explain the acceleration and expansion of the universe? Visible matter seems to account for barely 4% of the mass of the universe. Dark mass offers some solution to account for the shape of the universe and the connection among galaxies and the rotation within them. But then if dark matter exists, how is it possible for the universe to have expanded? Would not the significant mass of the original primordial atom from which the universe sprang been so great that gravity would prevent expansion? Would it not be like matter connected to a rubber band? Wouldn't the matter expand out only so far before it was snapped back by gravity?

Einstein was perplexed by this problem. He tried to account for the gravitational effects of the mass of the universe by proposing a cosmological constant. This constant would counteract gravity, making possible an expansion of the universe. Although Einstein at one point would repudiate his cosmological constant, what he was envisioning was some force or energy that was not visible, also able to account for the expansion of the universe. This concept, related to dark matter, is called dark energy.

Finally, the concepts of dark matter and dark energy are connected to the Big Bang theory. This theory suggests that the universe would infinitely expand after the Big Bang. But that expansion is contingent upon the energy and matter that exists in the universe. If the total density is low gravity will not exert a force to prevent this infinite cosmological expansion. Conversely, if the density is great enough then at some point the universe will stretch out as far as it can, only to be pulled back by gravity. If the latter occurs, this contraction is referred to as the Big Crunch. Whether the universe will infinitely expand or eventually contract is contingent upon the overall amount of matter and energy that exists – both visible and dark. In other words, depending on the density parameter of the universe and what the Hubble' constant actually is, that will determine the ultimate fate of the universe.

The concepts of dark energy and dark matter are controversial. Actual proof and capture of antimatter and dark energy have proved elusive. Moreover, the origin of both dark matter and dark energy is in debate. Astronomers and physicists have classified matter into baryonic and dark matter. The former is visible matter, what we see. It is characterized by being composed of some elementary subatomic particles, with the most common baryonic ones being protons and neutrons. Dark matter has a different particle makeup, and its exact composition is in dispute. But efforts to explain mass in the universe are presently directed at finding the Higgs boson, a particle physicists speculate will, once found, explain the origins of mass in the universe. Efforts to locate the "God particle," as it is known in pop culture, are now occurring at the Large Hadron Collider built by the European Organization for Nuclear Research (CERN) and which became operational in 2009.

When all the mass and energy calculated to be in the universe is added up, approximately 4% is visible matter. Of the remaining, 22% is calculated to be dark matter and 74% dark energy.

So how does all this connect back to the Andromeda Galaxy? Vera Rubin in a 2006 *Physics Today* article, "Seeing Dark Matter in the Andromeda Galaxy," talked about how her research on this galaxy and its rotation provided crucial data regarding the existence of dark matter and eventually dark energy. Efforts to explain M31's rotation back in the 1930s with Hubble and Zwicky led to the discovery of the dark side of the universe, and the Andromeda Galaxy continues to occupy an important role in explanations of galactic and cosmological evolution and formation.

Andromeda, Supernovae, Black Holes, and Other Cosmological Discoveries

The importance of M31 does not end with Hubble, Einstein, and an accelerating universe. After World War II the Andromeda Galaxy continued to be an object of curiosity, studied extensively to provide clues to many other astrophysics puzzles. The Andromeda Galaxy has also been studied to understand supernova, black holes, and a host of other phenomena.

Novae and the Andromeda Galaxy have been associated with one another every since S Andromedae, or SN1885A. Originally discovered by Irish astronomer Isaac Ward, the bright object in the Andromeda constellation became central to the island-universe debate. Explaining its brightness was a big part of the Great Debate in 1920 between Curtis and Shapely. Was the brightness an anomaly that otherwise demonstrated how far away Andromeda was because all the other stars in the nebula were faint and therefore it had to be a distinct galaxy (Curtis)? Or was its brightness proof that the nebula could not be so far away and therefore M31 was really part of the Milky Way? The curiosity of the supernova troubled astronomers into the early twentieth century. Additionally, although Edwin Hubble primarily relied upon Cepheid variables to help him track the distance to M31, he also tried to use supernovae as benchmarks. However, their brightness confused his analysis, throwing off his estimate of the distance to the Andromeda Galaxy.

What are supernovae? They are stars at the end of their lives. For stars whose masses are less than eight times that of the Sun, as they age they eventually burn out their core, and the remaining star settles into becoming a white dwarf. White dwarfs cool to a temperature of approximately 10,000 K and collapse until they about the size of Earth. They are dense, with a tablespoon-size piece weighing several tons. White dwarfs represent a calm end for stars.

But for stars whose masses are greater than eight times that of the Sun something else happens. As the star ages its core burns progressively hotter because the star is contracting upon itself. At some point when the core contracts sufficiently its temperature reaches hundreds of millions of Kelvin. The photons or energy is powerful enough to create a different subatomic particle, a neutrino. Neutrinos try to escape the star, and that requires it to either burn more of the core, contract, or both. At some point the nuclear reactions produce an iron core. At this point contraction is even more rapid, and the temperature jumps to 5×10^8 K or greater. What then occurs is a rapid contraction to the core followed by an even more rapid rebound off the core. The rebound off the core essentially explodes away the outer shell of the star, expelling energy and light in amounts that are hundreds of times greater than the Sun has emitted in its entire 4–5-billion-year history. The result is a super bright star – a supernova. This is called a Type II supernova. The Crab Nebula, M1, is an example of the remnants of a supernova.

There are also Type I supernovae. These are exploding stars, often white dwarfs, displaying spectrum lines of ionized silicon that is produced as a result of carbon being burned in a star. The reason why a Type I supernova occurs in a binary system is that its gravitational force may capture matter from another nearby star, providing additional fuel for an eventual heating and explosion. An example of a Type I supernova is the one the Tycho Brahe observed in 1572. This is called a Type Ia supernova. Astronomers have come up with several classes of Type I supernovae. What is important about a Type Ia is that it rapidly brightens and that all stars that go this route have approximately the same luminosities.

224 The Andromeda Galaxy and the Rise of Modern Astronomy

FIGURE 9.10 SN 1987A (*lower right*) in the large magellanic cloud (Courtesy of the Anglo-Australian University).

Calculating the apparent brightness and its period makes it possible to use Type Ia supernovae to calculate stellar distances.

Astronomers now have a theory to explain supernovae and why they become so bright and expel so much energy. They refer to the problems of Hubble, Curtis, and Shapely when seeking to account for the brightnesses of stars. So far only one supernova has been actually recorded in the Andromeda Galaxy, but supernovae have been recorded in other Local Group galaxies such as in the Large Magellanic Cloud, SN 1987a.

Black Holes and Gravity

Supernovae and white dwarfs represent two possible endings for stars. But another possibility exists – black holes. Recall Einstein's general theory of relativity. It states that gravity is a force that has the capacity to bend space-time. In the same way that gravity can bend light, it has the same capacity to do that with space-time.

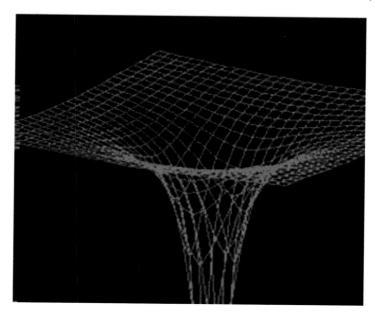

FIGURE 9.11 Depiction of a black hole distorting gravity and space-time.

In the general theory of relativity space-time is depicted as a plane with gravity as a sinkhole that curves around a massive object exerting force upon it.

Massive objects have the capacity to alter space-time and distort light. Moreover if an object is massive enough it might exert so much force that it even prevents light from escaping from its pull. Thus, when viewed the object would appear to be visually black, creating a black hole. What could cause such an object to exist?

Smaller stars that are not too massive burn out, die, and turn into white dwarfs. But some stars are more massive, more than $1.4\times$ the mass of the Sun, and the gravitational pull of their own weight turns them into neutron stars. Neutron stars are very small, compact and dense stars composed almost singularly of neutrons. The idea of such a type of star was originally proposed by Fritz Zwicky and Walter Baade back in the 1930s. They envisioned a massive dying star collapsing on itself. The star is named after the neutron, a particle formed when protons and electrons fuse together under high pressure.

Neutron stars often begin rotating rapidly as they collapse. As they rotate they shoot out a radiation beam along their magnetic axis, creating what are known as pulsars. One of the most famous pulsars ever detected is the star at the center of M1, the Crab Nebula. It was the death of this star that, when it exploded, produced the supernova of 1054, eventually yielding the nebula and the pulsar.

Some stars are even more massive, though. When their mass is greater than 3× that of the Sun, their death and collapse does not produce either a white dwarf or a neutron star. Instead, the mass of the star is so great that it continues to collapse upon itself. The space-time around the star gets bent, and eventually light around it disappears. What has been created is a black hole.

Astronomers have located thousands of black holes, but how? They cannot be seen because they eat light, thereby preventing them from being seen. However, they literally leave black holes or blotches in space, allowing astronomers to "see" these holes. But more importantly, other forms of observation, such as by radio or X-ray telescopes, allow astronomers to locate them across the sky.

Are black holes truly black? Not necessarily, according to astrophysicist Stephen J. Hawking (1942–). As stars collapse upon themselves they form a point of singularity, the black hole center itself where space-time is literally bent into itself. Around the hole is an event horizon, or the place where the escape velocity (the speed necessary to escape the gravitational pull of the black hole) equals the speed of light. Theoretically the event horizon traps all energy, including light, making it impossible for any to escape. This is what creates the black hole appearance – the inability of light to escape from its clutches. Thus the general law of relativity would dictate that black holes are one way passages; what goes in never comes out.

Yet, according to Hawking, Einstein did not take into consideration the principles of quantum mechanics. Quantum mechanics was developed in the early part of the twentieth century to explain the behavior of atomic and subatomic particles. Werner Heisenberg (1901–1976) stated that it was impossible to have an accurate description of the velocity and location of a subatomic particle. This claim is known as the Heisenberg uncertainty principle.

This principle includes the concept of virtual pairs – for every particle in space an antiparticle also exists. Antiparticles are like particles but with opposite charges. Particles and antiparticles are constantly being created and destroyed.

When a black hole eats a particle it leaves its antiparticle outside, forcing the latter to become a real particle. When it does, some of the black hole's energy (consistent with $E = mc^2$) will be converted to matter. Thus, the consumption of some matter, according to quantum mechanics, will convert antiparticles into particles. The result is that black holes may leak.

Why is this important? Some have speculated that black holes, in distorting space-time, have the potential of creating wormholes to other parts of the universe. They would curve the universe and space-time in such a way as to create bridges or shortcuts across the universe. Perhaps their bending of the universe and leakage means that one could use black holes as passages for time travel.

Science fiction writers see black holes as potentially useful for time travel and for crossing the universe at rapid speeds. Little evidence exists that this would be possible, though. The gravitational crush of a black hole would destroy anything that enters

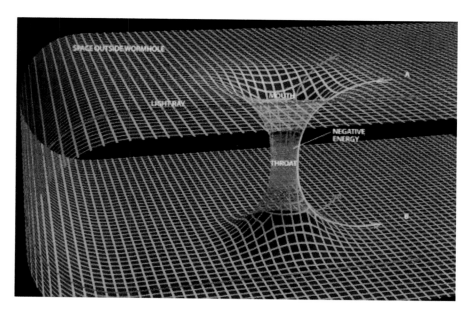

FIGURE 9.12 Black holes as wormholes to other parts of the universe.

it. Additionally, Hawking has argued that the proof against time travel is that we do not have any visitors from the future. However, this latter argument is not as good as it sounds. If the future has yet to be experienced then it would be impossible for us to receive visitors from it.

Galactic Black Holes

Black holes can also take on galactic dimensions – literally. When galaxies are formed they are originally a mass of gas that eventually collapses. Some of this gas at the center may simply collapse upon itself, forming a massive black hole at the center of the galaxy. With galaxies such as the Milky Way, which has an estimated mass of 2×10^{11} suns, or the Andromeda Galaxy, having a mass of 3.2×10^{11} suns, the possibility of there being a supermassive black hole at their cores is high.

Of course, supermassive black holes cannot be seen visually, but evidence of their existence abounds. The center of the Milky

FIGURE 9.13 Sagittarius A* captured by Chandra.

Andromeda, Cosmology, and Post-World War II Astronomy 229

Way is located in the constellation Sagittarius. The central region is obscured by stars, thereby making it difficult to see it. However, astronomers call the center of the Milky Way Sagittarius A*.

Visually, the area around Sagittarius A* looks simply like a crowded area of stars. However, this area has been studied and photographed by many orbital observatories. One picture is from the Chandra telescope, which captures X-ray images. What appears in the image is a bright white area. Some might think this is the image of the supermassive black hole, but it is not. The core around the black hole is densely packed with stars. These stars are in tight rotation around the black hole. A Keck Observatory image of Sagittarius A* imposes estimated or predicted orbits for the stars revolving around it. This motion and revolution around the black hole has served as proof that such holes exist.

But is the Milky Way unique in possessing a supermassive black hole? No. Astronomers have also detected black holes at the center of other galaxies, including M31. In January 2000, Chandra also captured an X-ray image of the Andromeda Galaxy's black hole. In this false color image the blue dot at the center represents what and where scientists believe the black hole is located. Revolving around it are stars in the central region. Just as with

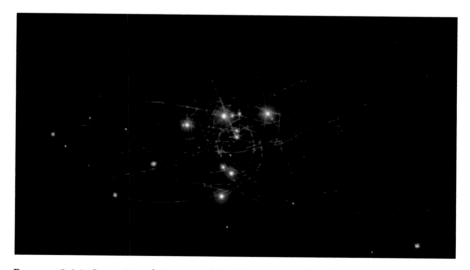

FIGURE 9.14 Stars in orbit around Sagittarius A* (Keck observatory).

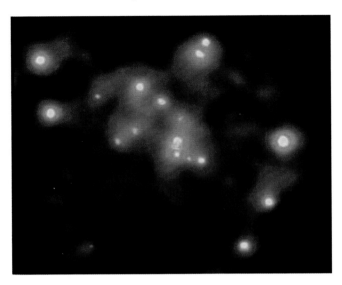

FIGURE 9.15 X-ray image of Andromeda's black hole (Chandra).

the Milky Way, Andromeda's black hole was confirmed indirectly by way of looking for rotational patterns around it. The study of M31's black hole, in conjunction with that of the Milky Way, has led to confirmation that many galaxies also have these objects at their centers, distorting space-time and gobbling up objects that fall within their gravitational forces.

Astronomers began searching for evidence of Andromeda's black hole beginning in the 1980s. Besides looking for the rotation of stars around it, the black hole itself should have revealed Doppler line shifts caused by distortions of its force. These Doppler shifts would likely reveal the orbital velocity around the core. John Kormendy and Ralph Bender in a 1999 article looked exactly for these effects. They found significant red- and blueshifts of stars in close rotation around the supposed core, providing evidence of some supermassive object distorting the paths of these stars. This was evidence of black hole. But in this 1999 paper something was found – a double nucleus at the heart of Andromeda Galaxy. They found this again by way of examining the distortions in stellar orbits around the core region. What did this mean? That the Andromeda Galaxy was the product of at least one other galaxy. Koremendy and Bender had observed the two cores that lied at the heart of the two galaxies that now made up M31 – evidence of galactic cannibalism.

Conclusion

Post World War II astronomy discovered literally that there was more to the universe than meets the eye. The discovery of black holes and dark energy and dark matter meant that the universe was more dense and crowded than thought. The region of the universe that the Andromeda Galaxy occupied hid a lot of secrets, and it was also a galaxy that seemed to be in flux, just as many of the ancient Greeks had thought. M31 came into existence by way of the Big Bang and the merger of other galaxies, produced by the forces of black holes, gravity, and the life and death of stars.

10. Astronomy and Andromeda at the Close of the Twentieth Century

Astronomy is a revolutionary science. The Copernican Revolution changed how humans saw the universe. So did the telescope, as did the invention of astrophotography and spectroscopy in the nineteenth century. Hubble's calculations of distance expanded the dimensions of the universe, Einstein's special theory of relativity decentralized it, and Lemaître's primordial atom gave it a new origin. Astronomy after World War II inherited all of these revolutions. By the later 1940s and early 1950s the universe was one created by the Big Bang, constantly expanding, billions of light-years in age and dimensions, and full of countless stars and galaxies beyond the Milky Way.

Beginning almost immediately after the Second World War, astronomy would experience changes that would expand knowledge about the universe even more. These changes were about size, location, and medium. Specifically, they included the building of bigger and bigger telescopes (size), space exploration and the launching of space satellites and observatories (location), and viewing the cosmos literally in new lights that began with radio astronomy and eventually expanded to other light frequencies beyond what the naked eye could see (medium). If astronomy was once a visual eye science centered on the Earth, post World War II astronomy changed that forever, opening up a new universe that both refined and redefined the cosmos.

Astronomical research after World War II branched off into many different directions as size, location, and medium changed. Yet the Andromeda Galaxy was central to many of the revolutions in astronomy that occurred during the second half of the twentieth century.

The Telescope Revolution

Since 1610, astronomers have sought to build bigger and more powerful telescopes into order to see further and better. In the first half of the twentieth century George Ellery Hale was a major force pushing for an expansion of visual telescopic astronomy. He was behind the establishment of the Yerkes Observatory in Wisconsin near Chicago in 1897. It housed what is still the largest refractor telescope in world, featuring a 40-in. lens. Hale also established the Mount Wilson observatory in California with a 60-in. reflector telescope that became operational in 1908, and then the 100-in. Hooker reflector that was the largest in the world from 1917 until 1948. The Hooker telescope was central to the research done by Edwin Hubble and to many of the galactic and astronomical discoveries leading up to World War II.

Hale was a dreamer. He wanted even a bigger telescope than the Hooker. Almost as soon as the Hooker 'scope' was completed Hale planned for a grander project – a 200-in. reflector to be located near Mount Wilson. The process for the planning and construction of this larger project began in 1928, when Hale secured a $6 million grant from the Rockefeller Foundation. Originally the new telescope was slated to be built at Mount Wilson, but the increasing light and atmospheric pollution from nearby Los Angeles made the site eventually seem unsuitable. Hale searched across the country including in Texas and Hawaii. Eventually he settled on 5,600-ft altitude of Mount Palomar, 100 miles southeast from Pasadena, California. Hale then bought 160 acres of land for the project.

The next problem was casting the 200-in. mirror. During 1936 to 1936 General Electric made an attempt to create the mirror out of fused quartz, but after spending $600,000 in unsuccessful attempts, they give up. Hale then turned to Corning Glass Works in New York. Corning proposed to make the mirror out of a new material – Pyrex – which had the advantage of expanding and cracking less. After one failure, they were successful in casting it.

While Corning was casting the mirror the details for the observatory design were undertaken during 1934–1936. Construction began after this, and the tricky task of moving the mirror and mounting it on the telescope commenced in the later 1930s. It took 16 days to move the mirror from New York to California by train. The train only moved 25 miles per hour. Once it arrived in California the mirror was ground and polished at Caltech; this took 11 years, from 1936 until 1947, with delays of a few years due to World War II. Finally in 1947–1948 the mirror was installed, and in 1948, before the telescope was fully operational, it was dedicated in memory of Hale, who had passed away in 1938. By 1949 the new 200-in. telescope at Mount Palomar became fully operational.

Walter Baade was among the first to use the new Mount Palomar telescope. He continued research he had begun at Mount Wilson, where only a few years earlier he had successfully resolved stars in the Andromeda Galaxy. It was at Mount Palomar that Baade undertook observations of M31 leading to important discoveries about stars, as well as offering important corrections to

FIGURE 10.1 Mount Palomar telescope.

FIGURE 10.2 Image of a starburst region of dwarf galaxy IC 10 taken with the Keck II telescope.

FIGURE 10.3 Keck I and Keck II observatories.

FIGURE 10.4 Gran telescopio canarias.

many of Hubble's conclusions, which are discussed later in this chapter.

However, bigger did not stop with Mount Palomar. Even larger ground-based telescopes have been constructed since 1948. Palomar remained the largest visual telescope in the world until 1975, when the Soviet Union brought on line BTA-6, a 238-in. reflector. After that even larger visual telescopes were constructed. Keck I and Keck II, constructed in 1993 and 1996, respectively, were telescopes of 400 in. each placed at the summit of Mauna Kea, Hawaii. Then in 2009 a 410 in. telescope became operational at the Gran Telescopio's Canarias Observatory in Chile.

Radio Astronomy

Post-World II astronomy was not all about bigger; it also transcended into new media or energy to observe. Historically astronomy has been a visual endeavor. Until 1610 astronomical observing was done with the naked eye. The invention of the telescope and Galileo's use of it to examine the sky literally magnified the universe but still astronomy involved observation of the heavens in visible light. The invention of photography eliminated the need to place an eye on the telescope, with a camera lens replacing it. Yet photography was still capturing images in the visible light spectrum. Even the discovery and employment of spectroscopy confined astronomical research to the visible spectrum. But visible light is only one small part of the electromagnetic spectrum. Other forms of energy are emitted by stars and other cosmological entities. For example, in 1912 Viktor Hess discovered in a high-altitude balloon experiment cosmic rays from space striking Earth. Thus, with the correct instruments perhaps these rays, and the objects that emit them, could be observed beyond what can be seen in the visible light.

Radio transmission for communication purposes clearly transformed the world in the first half of the twentieth century. The ability to broadcast live speeches, report news as it happened, or simply to speak to millions of individuals across the planet had an impact upon the world similar to the effect the Internet and the worldwide web had at the end of the twentieth and beginning of the twenty-first centuries. But as anyone who has turned on radios hears, there is often a lot of noise or static on the radio, especially with AM stations. Oftentimes the source of the interference is obvious. Drive under or near electric power lines, for example, and the static seems to overwhelm radio reception, yielding crackling and other noises. But sometimes the source is in doubt. Efforts to locate the sources of interference with radio transmissions led to the discovery of a new branch of astronomy.

In the 1930s Bell Laboratories, part of the engineering and research arm of American Telephone and Telegraph (AT&T) in New Jersey, was one of the leading edge scientific centers in the United States. Around 1931 one of its engineers, Karl Jansky (1905–1950), was seeking to troubleshoot the source of interference with a new transatlantic radio link.

Jansky built a radio antenna and aimed it at various objects. He found that thunderstorms and lightning caused static. He then aimed it at the Sun, detecting one source of interference, but he also found that there were other unknown causes. He detected that the interference seemed to peak every 23 h, 56 min – the actual length of a day on Earth. Eventually he determined that a source of static was coming from a point in the constellation of Sagittarius, an area considered the center of the Milky Way Galaxy.

FIGURE 10.5 Karl Jansky.

FIGURE 10.6 Karl Jansky's "radio telescope".

He concluded that the radio static he was picking up had an astronomical source. In 1933 he published a paper reporting his results and eventually urged Bell Labs to build a bigger antenna to capture the astronomical radio waves, but AT&T declined.

One person who read of Jansky's discovery was an amateur astronomer Grote Reber. He was an Illinois radio engineer, and in 1936 he built a simple radio telescope. It was a 31 ft (10-m)

FIGURE 10.7 Grote Reber and his radio telescope.

diameter dish with a receiver. Until 1944 he scanned the skies with his radio telescope, gathering transmissions at 1.9 m and 0.93 m wavelengths. In 1944 he turned his radio telescope to the Andromeda Galaxy, but he was unsuccessful with his instruments in isolating a transmission.

The first successful radio map of M31 was otherwise reported in 1951 by Robert Hanbury Brown and Cyril Hazard at the Jodrell Bank Observatory, located in Lower Wirthington, Cheshire, England. Over a period of 3 months during the fall of 1980 they scanned the skies with their radio telescope, eventually detecting a source coming from M31. In fact, M31 was the first object they examined once the Jodrell Bank telescope was ready for observing.

Their discovery was significant because it confirmed that radio waves and transmissions were not unique to the Milky Way but instead included other galactic centers. Among the conclusions they reached in a 1951 article were that the radio frequency

Astronomy and Andromeda at the Close 241

emissions from M31 were more intense than the Milky Way Galaxy (because of the greater mass of the former) but that overall the emissions from both were similar. Hanbury Brown and Hazard also surveyed other celestial objects, such as Tycho Brahe's supernova remnant. It was really their work and the research at the Jodrell Bank that ushered in the era of radio astronomy. It was this research on Andromeda that paved the way for additional

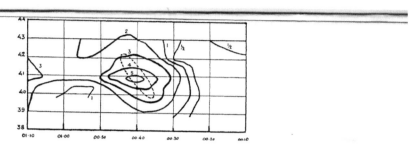

FIGURE 10.8 Hanbury Brown and Hazard radio telescope map of the Andromeda galaxy.

FIGURE 10.9 Lovell telescope at the Jodrell Bank observatory.

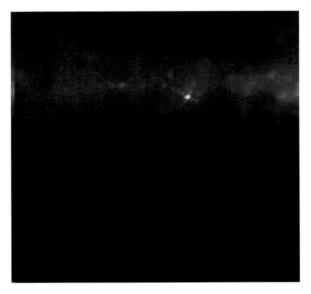

FIGURE 10.10 M31 radio image at 0408 MHz.

research, concluding that the Milky Way had a similar structure to that of M31. Subsequently, others such as English astronomer John Baldwin concluded in a 1958 article that the radio intensity of objects in M31 and the Milky Way matched their brightness in visible light – bright objects were also noisy.

The Andromeda Galaxy is quite different when viewed or recorded through a radio telescope. For example, the central core of M31 at the frequency of 0408 MHz looks quite different from visible light depictions.

In the same way that the size and light-gathering power of visible light telescopes has constantly increased, so has the power of radio telescopes. Today, some of the radio telescopes far eclipse the size of the ones used by Jansky, Reber, and Hanbury Brown and Hazard. Among the largest of the radio telescopes on Earth are the Very Large Array in New Mexico and the famous Arecibo Observatory in Puerto Rico. Among the major discoveries of the latter are those confirming the period of the planet Mercury and revealing new information about the Crab Nebula.

Astronomy and Andromeda at the Close 243

FIGURE 10.11 VLR in New Mexico.

FIGURE 10.12 Arecibo observatory.

What does the use of radio astronomy yield in terms of knowledge about the universe? For one thing, it expands the range of energy emitted from objects that can be recorded and assessed. Since visible light is only one small range of energy that can be detected or seen, the use of radio astronomy expands knowledge about objects by giving scientists a broader range of ways to observe. Second, because stellar dust and clouds often obscure regions of the sky from visible light observation, one can use radio telescopes to "look through" clouds. In the case of examining the galactic center of the Milky Way, radio astronomy opens up new avenues for observation. Third, the use of radio astronomy increases potential times and opportunity for observation. For the most part, visible light telescopic observation of stars and distant galaxies must occur at night. One can use radio telescopes during the day to observe. The Sun might interfere with some observations, but in theory a powerful radio telescope can be turned on during the day to observe the universe. Finally, radio astronomy suggested that if radio waves could be recorded, other instruments might be devised to record other forms of energy emitted from objects in the sky. Eventually, tools to capture infrared, ultraviolent, and even X-rays were developed to do that and observe M31.

Radio astronomy was a major revolution in astronomy. Its use, for example, led to the discovery in 1964–1965 of the microwave background radiation. Discovery of this radiation was critical confirmation of the Big Bang theory. Arguments by Fred Hoyle and others in 1948 about this theory led them to hypothesize that if a Big Bang did occur there would have been a leftover radiation background "noise" from early on after the universe was created. The discovery of what is now called the Cosmic Microwave Background (CMB) by Robert Dicke and others provided empirical evidence of the Big Bang that visible light astronomy could not.

Astronomy in the Space Age

Astronomy and human history forever changed on October 4, 1957. That is the day the Soviet Union successfully launched *Sputnik I* into space and a satellite was placed into orbit around Earth. That event began the space race and led to the United States placing *Explorer 1* into orbit on January 31, 1958.

The race to space became a surrogate battlefield for the Cold War between the USSR and the United States, symbolizing a test of superiority between communism and democracy. The importance of the space race was underscored by President John F. Kennedy in a May 25, 1961, speech to Congress when he stated: "First, I believe that this nation should commit itself to achieving the goal, before this decade is out, of landing a man on the moon and returning him safely to the Earth. No single space project in this period will be more impressive to mankind, or more important for the long-range exploration of space; and none will be so difficult or expensive to accomplish."

The race to the Moon had important political repercussions, and the United States won, placing *Apollo 11* on the Sea of Tranquility on July 20, 1969. But the race to space had not only political implications but scientific and astronomical ones, too. It allowed humans to transcend Earth and observe many of the planets up close. Satellites went to all of the major planets, either orbiting or landing on them. Missions to asteroids, comets, to the Sun, and even to the outer limits of the Solar System have all occurred. All of these missions, either by NASA, the Soviet Union, Russia, China, or the European Space Agency, had changed the location of astronomical viewing. No longer were humans required to view the cosmos from the surface of Earth, far distant from objects ancients could only gaze at and contemplate.

Clearly the limits of up close examination for now have been confined to the Solar System. It takes years to send probes and missions to Jupiter, Saturn, or even Mars. At present the technology does not exist to send a mission to another solar system. Even using the fastest rockets now designed by space technology, Proxima Centuri, the closest star to Earth at 4.2 light-years, would take nearly 72,000 years to reach. For example, the Saturn V rocket that took humans to the Moon flew at nearly 25,000 miles per hour. In August, 2011, NASA launched the Juno probe to visit Jupiter. The spacecraft will travel at nearly 37,000 mph (58,000 kph) to the planet. It will take nearly 5 years to reach Jupiter. It takes light 48 min to travel from Earth to Jupiter.

At present, interstellar, let alone intergalactic, space probes are not feasible. This means that one should not expect a space mission to the Andromeda Galaxy in the foreseeable future. Yet that does not mean that the race to space has not been important to astronomical research for distant objects such as M31. Instead, NASA and other space agencies have launched orbital observatories into space, allowing viewing to occur without the interference of the distorting effects of Earth's atmosphere. Among the most famous of the orbital observatories is the Hubble Telescope, inserted into space by NASA and the space shuttle in 1990. This telescope has provided striking images of the universe that provide clarity unmatched on Earth.

Hubble's primary telescope is visible light. But it has instruments that record more than that. In the same way that Jansky's radio telescope allowed for astronomy to transcend the narrow spectrum of visible light to capture radio frequencies from objects in the skies, there are many other frequencies that can be captured and recorded with the appropriate instruments. Hubble is equipped with instruments that allow for the capture of near-infrared and ultraviolet light. Examining the skies beyond the frequencies of visible light are much the same as they were with radio astronomy. They allow for examination of objects in other media and energy spectra, giving different and wider glimpses of the cosmos than can be been seen with the eye, even with an optical telescope. Stellar or intergalactic dust and glare might make it difficult to see some objects in visible light, but viewed in infrared or ultraviolet layers of clouds are peeled away, providing visibility to the previously invisible.

Now combine the capacity to view the sky in different energy spectra with the advantages of escaping Earth's atmosphere. This is the advantage of orbital observatories. New media and a new location can be combined to produce images of space, including M31, that transcend visible light images found on Earth.

Since the 1960s NASA and other space agencies have placed into orbit scores of observatories that can capture a range of energy spectra. Among the major NASA observatories are:

Name	Launched	Type of telescope
Compton gamma ray observatory	1990	Gamma ray
Hubble	1990	Visible, ultraviolet, near infrared
Rosat	1990	X-ray
Chandra	1990	X-ray
Spitzer	2003	Infrared
GLAST (Gamma Ray Large Area Telescope)	2003	Gamma-ray
GAXEX	2003	Galactic ultraviolet and spectroscopic

In addition to these observatories that have more general missions to explore space, some are devoted to specific projects, such as looking for exoplanets, exploring the Sun, or examining the microwave background radiation. All of these observatories have dramatically contributed to astronomical knowledge, including that of the Andromeda Galaxy.

Beyond NASA, many other countries and space agencies have also placed observatories or satellites into space. During the Cold War the Soviet Union actively explored space with probes to planets along with various types of telescopes for viewing the cosmos. Since then Russia has remained committed to space exploration. The European Space Agency (ESA) has also undertaken many missions, most famously the Cassini-Huygens mission that began in 2005. It involved exploration of Saturn and the landing of a probe on the planet's largest moon Titan. The ESA has also cooperated with NASA on many missions. China, France, Spain, and many other countries have also worked together to explore space by sponsoring or developing orbital telescopes to explore the universe in various energy spectrums.

The Andromeda Galaxy Across the Spectrum

The Andromeda Galaxy has been explored and studied by many of the orbital observatories in many different lights. Let's begin with a classic visible light image of M31 as captured by the Hubble Telescope.

Additional images either of the Andromeda Galaxy or its core reveal different views of it.

Each of these pictures reveals or depicts M31 in a different light, literally. As the different instruments to collect UV, gamma rays, infrared, or energy sources have been developed, the Andromeda Galaxy inevitably has been one of the first objects that astronomers have examined with their new tools and technologies. They have done so because of its proximity to Earth and the Milky Way and for the simple reason that it remains a central object of curiosity, much in the way it was one of the objects in the sky that the ancients examined. The new technologies and instruments peel

FIGURE 10.13 M31 image from the Hubble telescope.

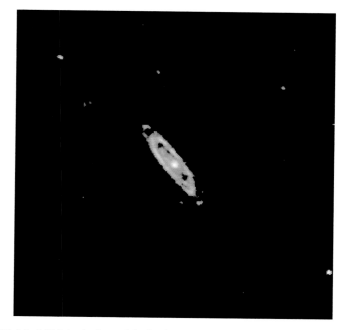

FIGURE 10.14 M31 in infrared light (IRAS 60 μm).

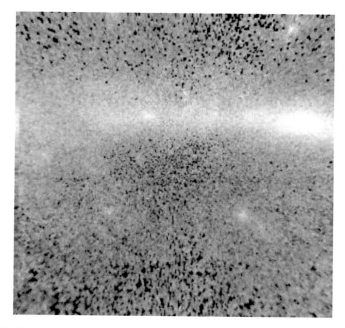

FIGURE 10.15 M31 in gamma rays (EGRET/NASA).

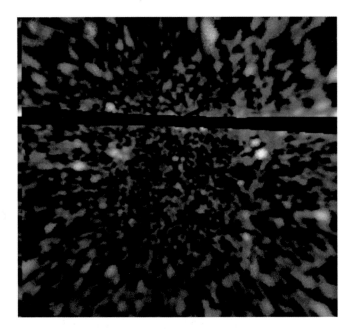

FIGURE 10.16 M 31 in X-rays.

FIGURE 10.17 M31 in near ultraviolet (GALEX).

back yet another new layer of M31, offering a new window into the galaxy and the cosmos.

Astronomical Revolutions and the Twenty-First Century

Two themes dominated astronomical research in the second half of the twentieth century and into the beginning of the twenty-first. The first was elaboration and refinement of the Big Bang model of the universe; the second were the revolutions that broadened how far we observe, how we observe, and where we observe. The Big Bang model articulates a theory about cosmological origins, and instrumentation and research since World War II has provided additional evidence that this model to explain the universe is basically correct. Clearly not all pieces of the puzzle have been addressed. There is still no answer to the basic question of why there is something as opposed to nothing. By that, why did the primordial atom exist and where did not come from and why did it suddenly go bang? Astronomers do not have answers to these questions. Nor do they have all the details about the universe mapped out. New curiosities and surprises take place daily, necessitating scientists to rethink basic theories. Yet these theories are produced in conjunction with the driving force of new observational tools that have provided astronomers new exploratory tools. Examination of the Andromeda Galaxy remains, and may always remain, at the center of efforts to understand the universe.

11. The Andromeda Galaxy into the Twenty-First Century and Beyond

The story of the Andromeda Galaxy tells the tale of the history and transformation of astronomy from myth to the science of astrophysics. Whereas the ancients and early modern astronomers saw M31 as a spot within a finite universe, the introduction of new technologies such as the telescope and spectroscopic analysis transformed our understanding of it from a nebula situated within the Milky Way into a distant island-universe. The study of the Andromeda Galaxy was at the apex of new theories in the twentieth century demonstrating that the universe was not a finite world composed of a single Milky Way Galaxy. Instead, through the works of Henrietta Leavitt, Edwin Hubble, Albert Einstein, Vera Rubin, Georges Lemaître, and others, the study of Andromeda was critical to the creation of an infinite universe of countless galaxies and vast distances, displacing humanity from its center into merely one point in its vastness.

Although historians of astronomy and science are familiar with how depiction of the universe changed, few have thought about how the study of the Andromeda Galaxy has been connected to the rise of astronomy as a modern science, and to our changing conceptions of the cosmos. Its study, and our advancement of astronomical knowledge, could not have been possible without giving M31 its long overlooked due.

What about the future of astronomical research and the Andromeda Galaxy? No doubt M31 will remain a central object of inquiry. Just think about a few possibilities.

The first is when it comes to the search for exoplanets. Prior to 1990 no planets had been detected outside of our Solar System.

FIGURE 11.1 The Andromeda galaxy.

Now it seems almost an everyday occurrence that another exoplanet has been discovered, with the count in the hundreds and into the thousands already. Many of these exoplanets are Jupiter sized, but as the instrumentation improves, smaller, super-Earths are being discovered. Moreover, solar systems such as the one we live in with multiple planets may not be unique. There could be millions of other stars with their own families of planets, perhaps some Earth-like, existing in the Milky Way Galaxy alone. Perhaps someday bigger and more powerful instruments will allow for discovery of exoplanets in the Andromeda Galaxy.

At one point no one thought that the stars of the Andromeda Nebula could be resolved; the future may bring discovery of an Andromeda Galaxy replete with many exoplanets and solar systems, perhaps some very similar to Earth and its Solar System. Demonstrating or discovering planets in the Andromeda Galaxy would then force the conclusion that perhaps many other galaxies, too, have planets and solar systems.

The Andromeda Galaxy into the Twenty-First Century 255

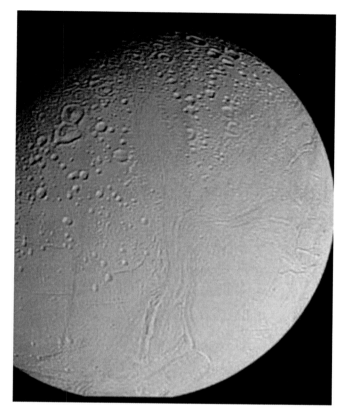

FIGURE 11.2 Enceladus.

Discovery of new planets and solar systems naturally then leads to another question: Will there be life on them? The study of astrobiology is the search for life beyond Earth. NASA has a team devoted to astrobiology. Exploration of Mars and discovery of water there in the first decade of the twenty-first century is pushing astronomers to speculation about life that might exist there. No, this is not the form of life that Giovanni Schiaprelli in the nineteenth century thought was behind the construction of the so-called Martian canals. Instead perhaps it is simple forms of microscopic life. Additionally, the Saturn moon Enceladus is full of water according to scientists, and due to the tidal forces of the Ringed Planet, the water may be liquid, capable of harboring life.

Moreover, although life as we know it on Earth is carbon-based, there are reasons to think that life may also be able to form around different chemical chains. The ESA probe to the moon Titan in the first decade of the twenty-first century revealed pools of liquid methane, prompting possibilities that maybe a different elixir can form the basis of life.

If there are billions and billions of stars in the sky, as astronomer Carl Sagan (1934–1996) was famous to utter in his PBS television series *Cosmos*, then why think that there is only one Earth-like planet in the entire cosmos? There could be many, and life might exist on most of them. The life may be something we can identify with in terms of being human-like, or perhaps it could be of a very different form. Future exploration for life in the universe, including the Andromeda Galaxy, might reveal life of some form there. Again, if life is there, then why not elsewhere in the universe?

Close Encounters of the Third Kind was a popular 1970s movie featuring actor Richard Dreyfuss who stumbles into a super-secret

FIGURE 11.3 Devils Tower, Wyoming.

NASA project revealing contact with extraterrestrial life. Dreyfuss eventually meets up with the aliens at Devils Tower, Wyoming.

Contact was made with them using a simple five-note melody – Re, Mi, Do, Do, So – made famous by the movie and repeated by many science fiction fans. Although perhaps not that simple, the SETI (or Search for Extra-Terrestrial Intelligence), asks amateur astronomers to use their computers to monitor noise from outer space with the hope and goal that they might detect signals from a distant planet or civilization. This is not the stuff of Erich von Däniken, whose 1968 *Chariot of the Gods!* and subsequent books wrote of UFOS and alien influences on ancient human cultures. Instead, this is a serious endeavor to find potential signals from intelligent life elsewhere, perhaps even within or from the Andromeda Galaxy.

Of course, were signals sent from the Andromeda Galaxy, traveling at the speed of light, it would have had to have left nearly 2.5 million years ago to reach us now. Whether such intelligent life existed then, or does now, or was or is communicating in ways we could understand, of course are very different questions.

Perhaps Einstein is wrong; maybe it is possible to time travel by exceeding the speed of light. Maybe even black holes, as some have speculated, can be used as special wormholes and shortcuts to travel across the universe in record time.

All of this is merely speculative now, and current science questions whether any of this is true. But what if all this is possible? What if someday science allows for intergalactic space travel at warp speed as depicted in *Star Trek* and many other science fiction shows, movies, and books? If someday space travel on this scale is possible, again no doubt missions to the Andromeda Galaxy will be a prime destination.

Longer term, the fate of Earth, Andromeda, and the Milky Way Galaxy is joined. Hubble was right to contend that M31 was a galaxy distinct from the Milky Way but wrong to think there is no connection between the two. They are part of a Local Network of galaxies. But unlike other local galaxies, M31 has a blueshift, indicating it is approaching the Milky Way. In 6 billion years the Milky Way and Andromeda will collide and merge into a supergalaxy. At that point Hubble will be wrong. Moreover, at that point, we no longer will have to travel to it to explore it; it will

FIGURE 11.4 The Andromeda and Milky Way galaxies merge.

have come to us. The final fate of the Andromeda Galaxy, should Earth survive 6 billion years, is that we will finally be up close and personal with an object that has inspired wonder since the beginning of humanity.

Appendix: Andromeda Facts

Andromeda Galaxy

260 Appendix: Andromeda Facts

Milky Way Galaxy

	Andromeda galaxy (M31)	Milky way galaxy
Hubble type	SB spiral	SB or SC spiral
Address and relatives	Largest member of local group of 30–40 galaxies	Smaller brother and member of local group of 30–40 galaxies
Mass	3.2×10^{11} suns	2×10^{11} suns
Diameter	220,000 light years	100,000 light years
Number of stars	4×10^{11}	2×10^{11}
Age	4.6 billion years	4.6 billion years
Rotational speed	220 km/s	250 km/s
Red shift	–300 km/s (moving toward earth)	NA
Number of globular clusters	300	150
Distance today	2.4 million light-years (720 KPC)	0
Distance in six billion years	0	Andromeda and the Milky Way collide and merge to form a super galaxy
Time for one rotation	225 million years	220 million years

References

Allen RH (1963) Star names, their lore and meaning. Dover, New York

Baade W (1944) The resolution of Messier 32, NGC 205, and the central region of the Andromeda nebula. Astrophys J 100:137

Baldwin JE (1958) The radio emission from the Galaxy and the Andromeda nebula. Comparison of the large-scale structure of the galactic system with that of other stellar systems. In: Roman NG (ed) Proceedings from IAU symposium no 5 held in Dublin, 2 Sept 1955. International Astronomical Union. Symposium no 5, Cambridge University Press, p 44

Barnard EE (1898) The great nebula of Andromeda. Astrophys J 8:226

Blumenberg H (1987) The genesis of the Copernican world. MIT Press, Cambridge

Burke BF, Fraham-Smith F (2002) An introduction to radio astronomy. Cambridge University Press, New York

Chant CA (1939) Our own island universe, the milky way. J Roy Astron Soc Can 33:72

Clerke AM (2003) A popular history of astronomy during the nineteenth century. Sattre Press, Decorah

Crosswell K (2001) The universe at midnight. The Free Press, New York

Curtis HD (1917) Novae in spiral nebulae and the island universe theory. Publ Astron Soc Pac 29:206–207

Curtis HD (1919) Modern theories of spiral nebulae. J Wash Acad Sci 9:217

Dufay J (1957) Galactic nebulae and interstellar matter. Philosophical Library, New York

Edge DO, Mulkay MJ (1976) Astronomy transformed: the emergence of radio astronomy in Britain. Wiley, New York

Elmegreen DM (1998) Galaxies and galactic structure. Prentice Hall, Upper Saddle River

Ely O (1909) Notes on an argument against an infinite universe. Popular Astron 17:209–213

Evans J (1998) The history and practice of ancient astronomy. Oxford University Press, New York

Ferris T (1988) Coming of age in the milky way. Perennial, New York
Frankfort H et al (1966) Before philosophy. Pelican Books, New York
Graves R (2003) The Greek myths, vol 1. The Folio Society, London
Greene B (2004) The fabric of the cosmos. Alfred Knopf, New York
Grego P, Mannion D (2010) Galileo and the 400 years of telescopic astronomy. Springer, New York
Hanbury BR, Hazard C (1951) Radio emission from the Andromeda nebula. Mon Not Roy Astron Soc 111:357
Hawking SJ (1998) A brief history of time. Bantam, New York
Hodge PW (1981) Atlas of the Andromeda galaxy. University of Washington Press, Seattle
Hodge WP (1992) The Andromeda galaxy. Kluwer, Boston
Hodierna GB (No date) Giovan Battista Hodierna: un precursore di Messier alla corte del primo dei 'Gattopardi. http://www.orsapa.it/hodierna/hodierna.htm. Accessed 14 Oct 2006
Holden ES (1881) Sir William Herschel: his life and works. Charles Scribner's Sons, New York
Hoskin M (1963) William Herschgel and the construction of the heavens. Oldbourne, London
Hoskin M (ed) (2000) The Cambridge illustrated history of astronomy. Cambridge University Press, Cambridge
Hough GW (1908) On an infinite universe. Popular Astron 16:461
Hubble E (1926) A spiral nebula as a stellar system, Messier 33. Astrophys J 63:236
Hubble E (1929) A spiral nebula as a stellar system, Messier 31. Astrophys J 69:103
Hubble E (1932) Nebulous objects in Messier 31 provisionally identified as globular clusters. Astrophys J 76:44
Hubble E (1934) The distribution of extra-galactic nebulae. Astrophys J 79:8
Hubble E (1935) Angular rotation of spiral nebulae. Astrophys J 81:334
Hubble E (1982) The realm of the nebulae. Yale University Press, New Haven
Hubble E, Humason ML (1931) The velocity-distance relationship among extra-galactic nebulae. Astrophys J 74:43
Hubble E, Tolman RC (1935) Two methods of investigating the nature of the nebular red-shift. Astrophys J 82:302
Huggins W, Miller WA (1864) On the spectra of some nebulae. Phil Trans Roy Soc Lon 154:437–444
Humason ML (1936) The apparent radial velocities of 100 extra-galactic nebulae. Astrophys J 83:10

Johnson G (2005) Miss Leavitt's stars: the untold story of the woman who discovered how to measure the universe. Atlas Books, New York

Jones KG (1969) Messier's nebulae and star clusters. Elsevier, New York

Jones MH, Lambourne RJA (2004) An introduction to galaxies and cosmologies. Cambridge University Press, Cambridge

Kant I (1969) Universal natural history and theory of the heavens. University of Michigan Press, Ann Arbor

Keenan P (1937) Studies of extra-galactic nebulae II. Astrophys J 85:325

King HC (2003) The history of the telescope. Dover, Mineola

Kirshner RP (2002) The extravagant universe. Princeton University Press, Princeton

Kormendy J, Bender R (1999) The double nucleus and central black hole of M31. Astrophys J 522:772–792

Koyré A (1957) From a closed world to an infinite universe. The Johns Hopkins University Press, Baltimore

Krupp EC (1983) Echoes of the ancient skies: the astronomy of lost civilizations. Harper & Row, New York

Kuhn TS (1957) The Copernican revolution: planetary astronomy in the development of western thought. Harvard University Press, Cambridge

Lavalle J et al (2006) Indirect search for dark matter in M31 with the CELESTE experiment. Astron Astrophys 450:1–8

Leavitt H (1912) Periods of 25 variable stars in the small magellanic cloud. Harv Coll Obs 173:1–3

Livio M (2011) Lost in translation: mystery of the missing text solved. Nature 49:171–173

Lockyer JN (1964) The dawn of astronomy: a study of temple worship and mythology of the ancient Egyptians. MIT Press, Cambridge

Lovejoy AO (1964) The great chain of being. Harvard University Press, Cambridge

Lundmark K (1924) The determination of curvature of space-time in de Sitter's world. Mon Not Roy Astron Soc 84:747–770

MacPherson H (1919) The problem of island universes. Observatory 42:329–334

McVittie GC (1937) The distribution of extra-galactic nebulae. Observatory 60:170

Melia F (2003) The black hole at the center of our galaxy. Princeton University Press, Princeton

Monck WHS (1909) The limits of the universe. J Roy Astron Soc Can 3:177

National Research Council (1921) The scale of the universe. Bulletin of the National Research Council (transcript of the Great Debate).

http://antwrp.gsfc.nasa.gov/htmltest/gifcity/cs_nrc.html. Accessed 29 Oct 2006

Oepik E (1922) An estimate of the distance of the Andromeda nebula. Astrophys J 55:406–410

Panek R (2011) The 4% universe: dark matter, dark energy, and the race to discover the rest of reality. Houghton Mifflin, New York

Pannekoek A (1961) A history of astronomy. George Allen & Unwin, London

Pease FG (1918) The rotation and radial velocity of the central part of the Andromeda nebula. Proc Natl Acad Sci USA 4(1):21–24

Phillipps S (2005) The structure and evolution of galaxies. Wiley, Chichester

Plato (1968) Republic. (trans: Francis MacDonald Cornford). Oxford University Press, New York

Redman RO, Shirley EG (1921) Photometry of the Andromeda nebula, M 31. Mon Not Roy Astron Soc 97:416

Reynolds JH (1921) The Andromeda nebula, M 33, and the nebucula major. Observatory 44:368–372

Reynolds JH (1938) Extra-galactic nebulae. Mon Not Roy Astron Soc 98:334

Robinson JM (1968) An introduction to early Greek philosophy. Houghton Mifflin, Boston

Rubin V (1995) A century of galaxy spectroscopy. Astrophys J 451:419–428

Rubin V (2006) Seeing dark matter in Andromeda galaxy. Phys Today 59:8

Saslaw WC (2000) The distribution of galaxies. Cambridge University Press, New York

Scheiner J (1899) On the spectrum of the great nebula in Andromeda. Astrophys J 9:149–150

Shapley H (1917) Note on the magnitude of Novae in spiral nebulae. Publ Astron Soc Pac 29(171):213

Shapley H (1919) On the existence of external galaxies. J Roy Astron Soc Can 13:438

Singh S (2004) Big bang. Fourth Estate, London

Slipher VM (1913) The radial velocity of the Andromeda nebula. Lowell Obs Bull 2:56

Spinrad H (2005) Galaxy format and evolution. Praxis, Chichester

Stebbins J, Whitford AE (1934) The diameter of Andromeda nebula. Proc Natl Acad Sci 20:93–98

Tannenbaum DG, Schultz D (1998) Inventors of ideas: an introduction to western political philosophy. St. Martin's Press, New York

Van Maanen A (1935) Internal motions of spiral nebulae. Astrophys J 81:336

Williams RC, Hiltner WA (1941) Dimensions and shape of the Andromeda nebula. Pub Obs Univ Mich 8:103–106

Wilson HC (1916) The light curve of T Andromedae. Ann Harv Coll Obs 80:135–145

Index

A
Adams, J.C., 21, 31, 89, 90, 182
Age of the universe, 204
Airy, G., 90
Alfarabi, 27
Algol A and B, 120
Almagest, 26, 40, 41
Al-Sufi, 39–40
Anaximander, 9, 10, 46, 49–51, 136
Anaximenes, 10, 14, 46, 49, 136
Ancient Chinese Astronomy, 10, 37–39, 123, 139, 184
Ancient Egyptian Astronomy, 2, 16, 27, 36, 37
Ancient Greeks, 2, 6, 8, 10–11, 13, 14, 19–23, 26, 36, 39, 45, 46, 51, 71, 117, 136, 138, 163, 231
Andromeda Galaxy
  distance, 60–61, 129–130, 176, 181, 215
  size, 60–61, 158, 178, 181
Andromeda Nebula, 3, 60, 63–67, 70, 90, 94, 98, 99, 102, 103, 106, 118, 122, 127, 134–155, 157–159, 162–165, 168, 169, 179–181, 191, 194–195, 211, 254
Antimatter, 221
Apollo 11, 245
Apollonius of Perga, 51–52
Aristotle, 6, 14, 22, 23, 27–29, 71
Asteroid Belt, 89, 97
Asteroids, 88, 245
Astrolabes, 36–37
Astrophotography, 96–99, 103, 106, 164, 233
Atoms, 10, 14, 46, 201–203, 220, 233, 251
Averroës, 27
Avicenna, 27

B
Baade, W., 180, 208–209, 211–212, 225, 235, 237
Babylonian astronomy, 35, 39
Bacon, F., 29–33, 73, 114
Barnard, E.E., 147–148
Bell Laboratories, 238, 239
Berlin Observatory, 90
Betelgeuse, 71, 72, 112–113, 116
Big Bang Theory, 8, 10, 17, 49, 73–74, 181–205, 207, 209, 217, 219, 221, 231, 233, 244, 251
Big Crunch, 10, 221
Binary stars, 120, 223
Bishop Raymond of Toledo, 27–28
Blackbody, 109–112
Blackbody radiation, 109, 111
Blackholes, 190, 222–231, 257
Blueshift, 128, 129, 146, 149, 155, 158, 164, 170, 180, 181, 190–193, 217–218, 230, 257
Bode, J., 87
Bode's Law, 88
Boeddicker, O., 143
Boltzman, L., 111, 112, 115
Brahe, T., 37, 52, 58–59, 75–77, 80, 123, 124, 184, 223, 241
Brown, R.H., 240–242
Busen, R., 100, 107

C
Cassiopeia, 2, 4, 19, 38, 59
Cepheid variables, 59, 120, 126, 127, 136, 163–165, 168, 181, 196, 211, 222
Cepheus, 19
Ceres, 88
Chandra, 123, 228–230, 247

267

Index

Christian Theology (and astronomy), 16–17, 26, 31, 61, 77
Clerke, A.C., 62
Close Encounters of the Third Kind, 256–257
CMB. *See* Cosmic microwave background (CMB)
COBE. *See* Cosmic background explorer (COBE)
Copernicus, N., 16, 17, 27, 28, 30–34, 41–43, 51–53, 55, 56, 67, 77–78, 91, 105, 106, 135–137, 148–149, 158, 169, 188
Cosmic background explorer (COBE), 202
Cosmic microwave background (CMB), 73–74, 202, 203, 220, 244
Cosmic Mystery, 77
Curtis, H., 149–151, 153, 154, 163, 170, 171, 176, 222, 224
Curved universe, 219

D
Daguerre, L., 96, 97
Dark energy, 204, 213–222, 231
Dark matter, 204, 213–222, 231
Deferents, 52
Democritus, 10, 11, 46
Descartes, R., 29, 30, 33, 56, 73
Dicke, R., 202, 244
Digges, T., 56, 57
Doppler, C., 128, 129, 190
Doppler effect, 128–130, 146, 189, 190
Doppler shift, 129, 130, 230
Draper, J.W., 97

E
Earth, 3, 22, 45, 69, 105, 135, 157, 181, 207
Easton, C., 142, 143
Einstein, A., 21, 84, 106, 132–133, 144, 187–190, 193–197, 201, 203, 205, 207, 219, 221, 222, 224, 226, 233, 253, 257
Einsteinian model of the universe, 187–190, 205
Electromagnetic spectrum, 108–117, 238
$E=mc^2$, 133, 187, 201, 203, 227
Enceladus, 60, 255
Energy, 101, 107–112, 115, 122–123, 133, 201–205, 213–223, 237, 238, 244, 246–248
Epicycles, 26–28, 30, 36–37, 41, 52, 53, 77, 79, 136, 137, 158, 213

Equant, 52–53
Eratosthenes, 51, 53
European Space Agency (ESA), 245, 247, 256
Exo-planets, 63, 95, 247, 253, 254
Expanding universe, 130, 182, 193–197, 205, 233

F
Flat universe, 185, 219, 220

G
Galactic formation, 181, 213–222
Galaxies, 1, 19, 45, 90, 105, 135, 157, 181, 207, 233, 253
Galileo, 16, 17, 35, 42, 43, 60, 64, 65, 75, 77–78, 85, 96, 97, 106, 136, 139–141, 185, 186, 237–238
Gamma rays, 108, 247–249
Gauss, K.F., 88
GAXEX, 247
General theory of relativity, 188–190, 193, 194, 196, 201, 207, 219, 224–226
Genesis, 13, 15, 25–26, 29, 45, 63, 182, 199–201
Geocentric model of the universe, 16, 26, 52, 74, 77, 82, 137
GLAST, 247
Globular clusters, 57, 65, 120, 122, 144, 145, 209, 211, 260
God, 8, 9, 12–17, 19–26, 29–35, 39, 45, 48–49, 57, 62, 63, 73, 77, 84, 91, 138, 139, 182, 199–200, 257
God's particle, 14, 221
Gran Telescopio Canarias Observatory, 237
Gravitational constant, 82, 220, 221
Gravity, 17, 31, 34, 80–84, 89, 118, 136, 138, 185, 188–190, 194, 202, 213–215, 219–221, 224–228, 231
Great Chain of Being, 15–17, 23, 31, 43, 45, 62, 63, 75, 91, 134, 179
Great debate, 124, 135–155, 157, 159, 163, 166, 171, 176, 179, 180, 222
Greek mythology, 19
Gregory, J., 59

H
Hale, G., 159–162, 234, 235
Halley, E., 43, 75, 86, 87

Halley's Comet, 43, 86, 87
Harvard Observatory, 126, 144, 159, 161
Hawking, S.J., 226–228
Heliocentric model of the universe, 30, 55–56, 74, 75, 133, 158, 169
Heraclitus, 13, 14, 48
Herschel, C., 125
Herschel, W., 44, 60–66, 75, 87, 95, 141, 142
Hertz, H., 108
Hertzsprung, E., 115, 116
Hertzsprung-Russell (H-R) diagram, 115, 116, 209
Hesiod, 9
Hess, V., 201, 238
Higgs-Boson particle (God's particle), 14, 221
Hipparchus, 51–52
Hodierna, G.B., 40
Hooker telescope, 160, 162, 234
Hoyle, F., 200, 244
Hubble constant, 130, 182, 195–199, 221
Hubble, E., 3, 19, 61, 64, 67, 70, 94, 127, 129, 147, 155, 157–181, 188, 191–194, 207–213, 222, 234, 253
Hubble's "sequence of nebular types", 177
Hubble Telescope, 4, 18, 65, 69, 114, 169, 183, 186, 187, 246, 248
Huggins, W., 102
Humason, M., 191, 192
Huygens, Christian, 59, 61, 85, 100, 247

I
Incan, 35
Infrared light, 18, 69, 171, 249
Islam, 27
Island-Universe, 61–64, 94, 121, 135–181, 195, 207, 208, 253

J
Jansky, K., 238, 239, 242, 246
Janssen, Z., 41
Jodrell Bank observatory, 240, 241
Joule, J., 111
Jupiter, 17, 21, 37, 42–43, 65, 69, 88, 89, 96, 130, 245–246, 254

K
Kant, I., 4, 6, 17, 64, 90–95, 104, 106, 141, 150, 152, 169, 170, 178, 212
Kant-Laplace Theory, 95, 142, 213

Kapteyn, J., 142–143
Keck observatory, 229, 236
Kennedy, J.F., 245
Kepler, J., 17, 24, 28, 52, 53, 76–80, 84, 86, 103, 105, 136–138
Kepler's laws of planetary motion, 17, 78–80
Kirchhoff's Laws, 100
Kirchoff, G., 100, 107
Kuiper Belt Region, 97

L
Laplace, P., 64, 90, 94, 95, 104, 106, 142, 212
Large magellanic cloud, 172, 174, 178, 224
Leavitt, H., 59, 118, 125–128, 136, 148, 155, 162–163, 165, 168, 211, 253
Le Gentil, G., 43
Lemaître, G., 193, 200–204, 233, 253
le Verrier, J., 90
Light, 3, 35, 57, 69, 106, 138, 162, 185, 207, 233, 257
Lippershey, H., 41–42
Local group, 172, 215, 224, 260
Lovejoy, A.O., 15
Lowell observatory, 76, 90, 146
Luminosity, 114–118, 120, 122, 124–128, 136, 148, 152, 153, 163, 165–168, 170, 171, 198, 211
Luther, M., 31, 200

M
M1 (Crab Nebula), 123, 183, 184, 223, 226
M13, 65
M27 (Dumbbell Nebula), 172, 175
M31, 3, 18, 19, 37–40, 43, 44, 58–67, 98, 99, 102, 103, 117, 122, 124, 127, 138, 142, 145–154, 157, 163–168, 170, 172, 178–181, 191, 192, 198, 205, 208–209, 211–212, 214, 215, 217–219, 222, 229–231, 235, 240–242, 244, 246, 248–251, 253, 257, 260
M33 (Triangulum galaxy), 90, 165, 167, 172, 173, 179
M42 (Orion Nebula), 139, 175, 210, 211
M45 (Pleiades), 139
M51, 142, 176
M81, 149
M99, 142

270 Index

M101, 152
M104 (Sombrero galaxy), 172, 174
Magellanic Clouds, 126
Malebranche, N., 29
Mars, 21, 26, 60, 63, 69, 74, 76–79, 88, 122, 132, 245, 255
Maxwell, J., 101, 106, 107, 130, 132
Mercury, 21, 37, 69, 88, 189, 242
Messier, C., 2–4, 43–44, 60–62, 64–66
Metius, J., 41, 42
Michelson-Morley experiment, 131, 132
Milky Way galaxy, 62, 66, 67, 138, 141, 145, 158–159, 171, 238, 240–241, 253, 254, 257, 260
Mimas, 60
Mira, 119–120
Moon, 4, 5, 13, 17, 21, 22, 27, 37, 39, 41–43, 46, 48, 49, 51–53, 55, 58, 60, 62, 65, 69, 81, 92, 96, 97, 99, 101, 105, 114, 182, 185, 245, 247, 255, 256
Mount Wilson observatory, 153, 159, 161, 163, 234

N
NASA, 123, 169, 174, 183, 186, 202, 203, 245–247, 249, 255, 257
Native-American Astronomy, 70–71
Nebulae (as unresolved stars), 45, 60, 64–67, 95, 102
Neptune, 42, 60, 89, 90, 136
Neutron stars, 225, 226
Newton, I., 17, 21, 24, 30, 34, 76, 80–86, 89, 90, 100, 101, 103, 105–107, 132, 135, 136, 138, 185, 188–190, 205, 214–215, 218
Newtonian Model of the Universe, 84
Newton's laws of motion, 34, 81, 83, 84, 89, 212
NGC 2403, 149
NGC 6946, 149
Nicholas of Cusa, 30, 55
Novum Organum, 31–32

O
OBAFGKM, 113–114
Occam's razor, 28, 30, 32
Olbers, H., 57, 88
Olbers paradox, 57, 182, 185, 187, 204–205
On the Revolution of Heavenly Spheres, 16, 30

Öpik, E., 166, 170
Orion, 4, 72, 113, 140, 211

P
Pallas, 88
Palomar Observatory, 162, 196
Parallax, 58, 59, 61, 77, 117, 118, 125, 127, 148
Parmenides, 14
Pegasus constellation, 1–4, 38, 39
Period-luminosity relationship, 163, 165, 166, 168
Petroglyphs, 37–38
Piazzi, G., 88
Planck, M., 202
Planck time, 202, 203
Plato, 11, 12, 14–15, 27, 29, 46
Population II Stars, 209, 211
Population I Stars, 209
Pre-Socratics, 12
Principia Mathematica, 17, 34, 85
Proctor, R., 122, 142
Proxima Centauri, 198, 199
Ptolemy, 2, 16, 25–27, 36–37, 40, 41, 43, 52–54, 56, 60, 62, 63, 80, 86–87, 117, 136–137, 141, 178–179, 205
Pulsars, 226
Pythagoras, 10–11, 138–139

R
Radio astronomy, 215, 233, 237–244
Red giants, 71, 73, 117, 119–120, 209
Redshift, 128–130, 146, 149, 152, 153, 158, 170, 180–205, 212–215
Revolution of the Heavenly Spheres, 16, 30
Ritchey, G.W., 149
Roberts, I., 58, 66, 98–100, 142, 145
Rosat, 247
Rosse, Lord, 44, 98, 142, 145
RR Lyrae stars, 144
Rubin, V., 214, 215, 222, 253
Russell, H.N., 115

S
Sagan, C., 256
Sagittarius A*, 144, 228, 229, 238
S Andromedae (SN 1885 A), 124, 149, 171, 222
Saturn, 21, 37, 42, 60, 65, 69, 88, 92, 245, 247, 255

Index

Schiaprelli, G., 255
Schwarzschild, K., 211
Scientific method, 32, 71, 73, 96, 135, 136
Search for extra-terrestrial intelligence (SETI), 257
Shapely, H., 127, 151–152, 154, 163, 165, 170, 171, 176, 179, 222, 224
Single-universe hypothesis, 94
Size of the universe (cosmological distance), 55, 58, 59, 61, 63, 64, 75, 127, 130, 148, 178, 181, 197, 200, 212
Slipher, V.M., 146–150, 170, 181, 190–191, 214
Small Magellanic Cloud, 126, 127, 148, 163, 165, 172, 215
SN 1572, 184
SN 1604, 184, 186
Socrates, 6–8, 72
Space-time, 188–191, 202, 219, 224–227, 230
Special theory of relativity, 17, 132, 133, 138, 188, 233
Spectral classification, 114, 115
Spectral lines, 100, 101, 114, 129, 146, 191, 192
Spectroscope, 102
Spectroscopy, 95, 99–103, 105, 106, 136, 138, 190, 233, 238
Spectrum, 85, 100–102, 113, 129, 144, 146, 214, 223, 238, 246–251
Spiral nebulae, 102, 146, 147, 149, 153
Spitzer telescope, 123, 171
Starry Messenger, 43
Star Trek, 257
Star types, 71, 113, 117, 225
Stefan-Boltzman blackbody law, 111
Stefan, J., 111, 112
Stellar Parallax, 61, 117, 125
Stonehenge, 23–24, 35, 36
Sun, 5, 8, 10, 13, 16, 17, 21, 22, 26, 29, 30, 35–37, 41–43, 46, 49, 51–53, 55, 58–60, 62, 66, 73, 78, 79, 81–89, 91–93, 99, 101, 102, 105–107, 109, 110, 112–116, 118–119, 122–123, 131, 137–139, 141–144, 157, 163, 165, 167, 168, 182, 188, 189, 198, 202, 209, 214, 223, 225, 226, 238, 244, 245, 247
Supernova, 59, 69, 74, 77, 78, 122–124, 145, 146, 148, 157, 163, 166, 171, 184, 185, 196, 198, 222–224, 226

T
Telescopes, type, 35, 114, 123, 247
Thales, 6, 7
Titan, 247, 256
Tolman, R.C., 192, 193
Tombaugh, C., 90, 97, 146
Type II supernovae, 122, 123, 223
Type I supernovae, 122, 223, 224

U
Ultraviolet light, 246, 248
Uranus, 87–90, 92, 136
Ussher, J., 77

V
Van Maanen, A., 179
Variable stars, 59, 118–125, 127, 128, 162–163, 196
Venus, 21, 37, 42, 60, 69, 86–88, 184
Virgo Cluster, 215
Visible light, 69, 106–117, 205, 207, 237–238, 242, 244, 246, 248
von Däniken, E., 257
von Fraunhofer, J., 100
Vulcan, 189

W
Ward, I., 122, 222
Wein, W., 110–111
White dwarfs, 115–117, 123, 223–226
Wien's law, 111, 122
William of Moerbeke, 27–28
William of Ockham, 28

X
X-rays, 108, 226, 229, 230, 244, 247, 250

Y
Yerkes Observatory, 159, 161, 234
Young, T., 100, 101, 106–108

Z
Zeno, 14
Zone of Avoidance, 154, 176
Zwicky, F., 218, 219, 222, 225

Printed by Publishers' Graphics LLC